EXPLORING MARS

Exploring Mars

Chronicles from a Decade of Discovery

Scott Hubbard

Foreword by Bill Nye

THE UNIVERSITY OF
ARIZONA PRESS

TUCSON

www.uapress.arizona.edu

Library of Congress Cataloging-in-Publication Data
Hubbard, Scott, 1948–
Exploring Mars : chronicles from a decade of discovery / Scott Hubbard.
p. cm.
Includes index.
ISBN 978-0-8165-2111-1 (hc. : alk. paper)
ISBN 978-0-8165-2896-7 (pbk. : alk. paper)
1. Mars (Planet)—Exploration. 2. Space flight to Mars—History. I. Title.
QB643.H83 2012
523.43072'3—dc23 2011036184

♻
Manufactured in the United States of America on acid-free, archival-quality paper containing a minimum of 30% post-consumer waste and processed chlorine free.

17 16 15 14 13 12 6 5 4 3 2 1

To Susan

Contents

Figures

Unless otherwise noted, all photos are provided by the author.

Plates

Color plates follow page 40.

Foreword

Despite its common nickname, it's not red so much as orange. Even then, someone has to carefully show the stargazing newcomer that the planet Mars really has a special color. It's close and bright; it has always been a special place for us. As you make your way through the history recorded in Dr. Hubbard's book, please notice the names. So many Earthlings were involved in learning about our mysterious neighbor.

It has that red or reddish hue, for one thing. And so perhaps the slightly pink planet was named for the God of War. Why else would it be red, if it were not for the blood of warriors' spilt in battle? This ancient theme carried to the last century in which dozens of science fiction writers imagined a planet controlled by beautiful, thoughtful, often hostile Martians, most of whom came to a bad end—or came here to ensure a bad end for Earthlings.

These imagined interplanetary conflicts were fueled by water—not real so much as imagined or inferred water. In 1877, when orbital motions conspired to bring Mars slightly closer to Earth than it is most of the time, the professional Italian astronomer Schiaparelli observed what he thought might be channels on the Martian surface. With the Italian word for such geological features being *canali*, the gentleman astronomer Percival Lowell in the early twentieth century went to great lengths to popularize the notion that there might be water-filled *canals* running in a network all over Mars. Science fiction stories often featured sailboats, Martian agriculture, and cool evenings by the water's edge. There's nothing like a worldwide drought to drive a fictitious Martian to drink. I mean to drink water from another planet—usually ours. And, there's nothing like the promise of water to drive a modern spacecraft-mission manager to develop instruments that can sniff for moisture or scrape for ice in an extremely cold, remote, and desolate world.

Based on my own experience with optical eyepieces of microscopes and telescopes, I've often wondered if people viewing Mars had overlain a reflection from the back of their own eyes onto the image in their telescopes. In a cosmic irony, they might have been interpreting the tiny pattern of blood vessels in their retinas as enormous patterns of canals on a world more than 50,000,000 kilometers away. Perhaps laughable to some of us, it is just one more problem for a scientist. How do you know when you've really discovered something new? My old professor Carl Sagan often pointed out that you have to get really close to know for sure what you're seeing. That's why for Hubbard et al. (me inclusive), Earthbound planet gazing is not enough; you need to send spacecraft to see the Martian world up close.

With modern views of Mars provided by the designers and builders of modern telescopes, we just don't see any canals or evidence of great civil-engineering works out there. But strange as it might seem at first, the scientists and engineers of the twenty-first century are driven by the same deep quest that drove the characters of science fiction in the last century. We all sought and seek water there.

This became Dr. Scott Hubbard's job, to harness the resources of a modern space agency to conceive, design, build, and launch spacecraft to look for signs of water and life, to be the Czar of Mars. Tasks difficult enough here on Earth become astonishingly difficult on a bitterly cold world hurtling through the icy blackness of space tens of millions of kilometers away.

If you think about it, it is almost incredible that we can see Mars at all, let alone aim for it and hit it with spacecraft. An object like Mars takes up a portion of sky about the same width as a thin thread held at arm's length. Don't take my word for this—try it. As bright as celestial objects may seem to us on a clear, dark night, they would be absolutely invisible if space itself were not so fantastically empty. Light goes beaming right through almost all of the volume of the universe. There is nothing to stop it. There is a lot of space out in space. When light, spherically moving out from the Sun, hits Mars, a tiny fraction of it, only about 0.02 percent, bounces back to your eye. But against nature's blackest background, there it is.

When you're trying to land a spacecraft near a certain mountain on Mars, this amazing emptiness becomes relevant on the scale of about one in a million. It's like hurling a baseball from a pitcher's mound in Tokyo and having it land in a catcher's mitt in Buenos Aires. That's just taking into account the geometry as measured with meters or miles. When you try such a feat while taking into account all the probabilities—combining the gambles, that is, the odds of all the things that can go wrong—it's not one in a million. The odds of success start to look more like one in a million in a million—more like one in a trillion. But Dr. Hubbard and his team did it—several times.

Ancient people were intrigued not only by what they imagined might be up there on the pale red planet but also by its remarkable, or just plain weird, motion. Mars occasionally, albeit astonishingly predictably, appears to move backwards, with *retrograde* motion, as seen against the other bright points of the heavens. Until Copernicus had the courage to publish his theory of planetary motion, along with remarkably accurate observations, no one could quite explain why Mars appears to move the way it does.

Mars doesn't move backwards, not at all. Instead, we Earthlings observe Mars from an inside track on a planetary raceway. Now and then, every 26 months or so, we catch up and pass our rusty neighbor on our inside lane. The Hubbard teams had to focus on this bit of orbital mechanics in every decision they made. With millions upon millions of bits of navigational data developed with more than a few million US dollars, you just can't afford to miss. As any resident of an ocean shore will tell you, the tide waits for no one. Neither does Mars. Time is money; so Scott Hubbard had to gather his gangs of scientists, engineers, and especially accountants, and get them to get along and then get going in time with the orbital rhythm of these two worlds.

For us Earthlings, Mars started out as a completely mysterious place. Our ancient myths of gods or bellicose civilizations are not especially relevant to our everyday lives here in the early twenty-first century. Neither are the more recent and equally flawed ideas of planetwide irrigation systems with their concomitant societies of placid beings. But Mars is more relevant to us than to any peoples who may have come before us—ever. With Scott's direction, we have discovered proof, real proof, that Mars was once a very wet world. Here on Earth, wherever—and I mean *wherever*—we find water, we find living things. Was it or even is it thus on Mars? The question is fundamental, deep within us. The answer, if we can find it, would address the deepest questions that we all, at some point in our lives, must come to terms with: Where did we come from? And are we alone? If you claim that you have never asked yourself these two questions, well, I don't believe you.

From time to time, in a museum, university, or rock shop, you may come across a real piece of Mars, a real rock or pebble that has made the trip across interplanetary space from Mars to our Earth. Observant people find these meteoric rocks in places like the stark white ice sheet near the Allen Hills of Antarctica. These might be stones or pebbles or grains. Geologists and astrobiologists examine these rocks ever so carefully. There are shock patterns frozen within what were once molten or plastic pieces of Martian crust. These striations are evidence of a powerful collision. There are minerals featuring chemicals and isotopes that match those on Mars. Occasionally, tiny bubbles of gas are liberated from a rock's insides with the help of carefully designed rock twister tools in vacuum chambers. The bubbles are microscopic volumes of the Martian atmosphere.

On other meteorites, we find complete amino acids, complicated chemical chains that exist within our cells. It's just that the ones in question came here from outer space. So is it possible that life, some living cells or microbes or some*things* (*sic*), made the trip from Mars to Earth and that we are all really descendants of Martians? It's not crazy; it's just plain astonishing. It may even be true.

No matter how you run your reasoning, if we were to find evidence of life on Mars, fossil microbial mats for example, it would change the world. If we were to find something or things still alive on Mars, it would knock this world on its heels. It would be akin to the discoveries of Copernicus or Galileo. It would change the way each and every one of us considers his or her place in space and our role in it.

It takes a man like Scott Hubbard not just to muse on these Martian mysteries but instead to pursue the truth. It takes a guy inspired by intrigues of his childhood to do the work of exploring this other world. As you read here, you will of course come across all manner of remarkable facts about Mars. You will be exposed to the absolutely astonishing complexity of the tasks Hubbard and his teammates undertook.

But quickly and throughout, you will learn that this business—real spaceships that are really sent to a real alien world—is not created by technicians devoid of emotion. Instead, the business of exploring Mars is conducted by people with a passion. It was their Mars Czar's job to keep them on track both with the orbits of these neighboring worlds and with each other. Getting there and learning what we've learned was not easy, but Dr. Hubbard will tell you how he pulled it off. He has, in no small way, changed the world, our world. Read on!

Bill Nye
14 September 2011

Abbreviations

APL	Applied Physics Laboratory at Johns Hopkins University
APXS	alpha-particle x-ray spectrometer
ARES	Astrobiological Reconnaissance and Elemental Surveyor
ASI	Agenzia Spaziale Italiana (Italian space agency)
CCAFS	Cape Canaveral Air Force Station (Florida)
ChemCam	remote chemical-sensing instruments for MSL
CheMin	x-ray diffraction/x-ray fluorescence instrument for MSL
CIC	Capital Investment Council
CNES	Centre National d'Études Spatiales (French space agency)
COFR	Certification of Flight Readiness
COMPLEX	Committee on Planetary and Lunar Exploration
CONTOUR	Comet Nucleus Tour
CRISM	compact reconnaissance imaging spectrometer for Mars
CSA	Canadian Space Agency
CTX	context camera
DAN	dynamic albedo of neutrons instrument for MSL
DARPA	Defense Advanced Research Projects Agency
DSN	Deep Space Network
EDL	entry, descent, and landing
ESA	European Space Agency
FPGA	field-programmable gate array
GRS	gamma-ray spectrometer
Hazcam	hazard-avoidance camera
HEDS	Human Exploration and Development of Space

HEND	high-energy neutron detector
HiRISE	High Resolution Imaging Science Experiment
IMEWG	International Mars Exploration Working Group
ISAD	icy soils acquisition device
JAXA	Japan Aerospace Exploration Agency
JPL	Jet Propulsion Laboratory (Pasadena, CA)
JSC	Johnson Space Center (Houston, TX)
KSC	Kennedy Space Center (Florida)
LIDAR	light detection and ranging
LP	Lunar Prospector
MAG/ER	magnetometer and electron reflectometer
MAHLI	Mars Hand Lens Imager for MSL
MARCI	Mars Color Imager
MARDI	Mars Descent Imager for MSL
MARIE	Mars Radiation Environment Experiment
MARSIS	Mars Advanced Radar for Subsurface and Ionosphere Sounding
MAST	Mars Ad Hoc Science Team
MastCam	video camera for MSL
MAV	Mars ascent vehicle
MAVEN	Mars Atmosphere and Volatile Evolution mission
MAX-C	Mars Astrobiology Explorer-Cacher
MCO	Mars Climate Orbiter
MCS	Mars Climate Sounder (a radiometer)
MDC	Mission Director's Center
MECA	microscopy, electrochemistry, and conductivity analyzer
MEDLI	entry, descent, and landing instrumentation for MSL
MEP	Mars Exploration Program
MER	Mars Exploration Rover
MESUR	Mars Environmental Survey
MET	Meteorological Station
MGR	Mars Geological Rover
MGS	Mars Global Surveyor
MI	microscopic imager
MIMOS	miniaturized Mössbauer spectrometer
Mini-TES	miniature thermal emission spectrometer
MO	Mars Observer
MOC	Mars Orbiter Camera
MOLA	Mars Orbiter Laser Altimeter

MPL	Mars Polar Lander
MR	Mars Relay
MRO	Mars Reconnaissance Orbiter
MSL	Mars Science Laboratory
MSO	Mars Surveyor Orbiter
NAC	NASA Advisory Council
NASA	National Aeronautics and Space Administration
NASA Ames	NASA Ames Research Center (Mountain View, CA)
NASA HQ	NASA Headquarters in D.C.
NASA Langley	NASA Langley Research Center
Navcam	navigation camera
NRC	National Research Council
OMB	Office of Management and Budget
OSS	Office of Space Science at NASA HQ
QMS	quadrupole mass spectrometer
RAD	radiation assessment detector
RAT	rock abrasion tool
REMS	Rover Environmental Monitoring Station for MSL
Roskosmos	commonly used name for the Russian Federal Space Agency
SAIC	Science Applications International Corporation
SAM	Sample Analysis at Mars instrument for MSL
SDT	Science Definition Team
ShaRad	shallow radar sounder
SIRTF	Space Infrared Telescope Facility
SNC meteorites	Meteorites composed of the shergottite, nakhlite, and chassigny classes, which share characteristics rare in known meteorites
SOMO	Space Operations Management Office (NASA)
SSI	surface stereo imager
TECP	thermal and electrical conductivity probe
TEGA	thermal and evolved gas analyzer
telecom	telecommunications network
TES	thermal emission spectrometer
THEMIS	thermal emission imaging system
TLS	tunable laser spectrometer
TRL	technology readiness level
USO/RS	ultrastable oscillator for Doppler measurements
WCL	wet chemistry lab

Acknowledgments

Many thanks to my editor in chief at the University of Arizona Press, Allyson Carter, for her patience in working with a first-time book author and to Debra Makay for her thorough and insightful text editing.

In the body of the book, it is clear that Dan Goldin, Ed Weiler, the late Earle Huckins, Firouz Naderi, Gentry Lee, the late Jim Martin, Jim Garvin, and Steve Isakowitz all played an extraordinary role in the development of the Mars Exploration Program.

There are many others, though, who deserve credit for working with me on this incredibly difficult but ultimately successful journey: I thank my key staff at NASA Headquarters: Dave Lavery, Mark Dahl, Ramon DePaula, Joe Parrish, George Tahu, Paul Hertz, Mike Meyer, Cathy Weitz, Bruce Betts, Joe Boyce, Voleak Roeum, Brandy Nguyen, and the late Jane Davis. At the NASA Ames Research Center I received enormous help over the years in writing, editing, and storytelling from Lisa Chu-Thielbar. At the Jet Propulsion Laboratory (JPL), the list of additional contributors is too long to be able to list here, but a few key people deserve special credit: Ed Stone; Charles Elachi; George Pace, the project manager for the 2001 *Odyssey*; and Pete Theisinger, the project manager for the 2003 rovers.

Our international partners were critical in arriving at a final mission queue: Enrico Flamini of Agenzia Spaziale Italiana (ASI), and Christian Cazeaux and Richard Bonneville of Centre National d'Études Spatiales (CNES). Finally, the science community provided invaluable input, comments, and critique during this entire process.

It is impossible to list the many dozens of other contributors, but a few individuals stand out: Steve Squyres, Mike Carr, Ron Greeley, and Mike Malin. I deeply thank all those listed and unlisted (the latter know who they are).

EXPLORING MARS

CHAPTER ONE

Mars Is Hard

On September 23, 1999, near 2 a.m. Pacific daylight time, the Mars Climate Orbiter, known as MCO, a National Aeronautics and Space Administration (NASA) spacecraft that had been launched nine months earlier from Cape Kennedy in Florida, was just arriving at the red planet. The control center at NASA's Jet Propulsion Laboratory (JPL) in Pasadena, California, was abuzz with excitement and expectation. The gratifying culmination of several years' work from one of the greatest space engineering centers ever established was nearly upon us. Concepts, designs, test programs, calculations, discussions, highly skilled teams carefully garbed in white suits, scientists, and engineers had all played their parts in ensuring mission success. Engineers sent the signal necessary to fire the spacecraft's engines and put the probe into orbit. The only thing left to do was wait the required ten minutes for the spacecraft to reemerge from its loop around the back of Mars. A simple signal would then be transmitted from the 2-meter-high (about 6-feet-high) craft the approximately 200 million kilometers (about 120 million miles) back to Earth, indicating that all was well.

The signal never arrived.

Deafening silence hardly describes the dismay and disbelief that filled the room. Over the next several weeks, possibilities were calculated and recalculated, theories were examined, searches were conducted with radio antennas listening from all over the world—to no avail. MCO had vanished.

As Mars and Earth each orbit the Sun on their set paths, they come in proper alignment for a launch every 26 months or so. A rocket must launch

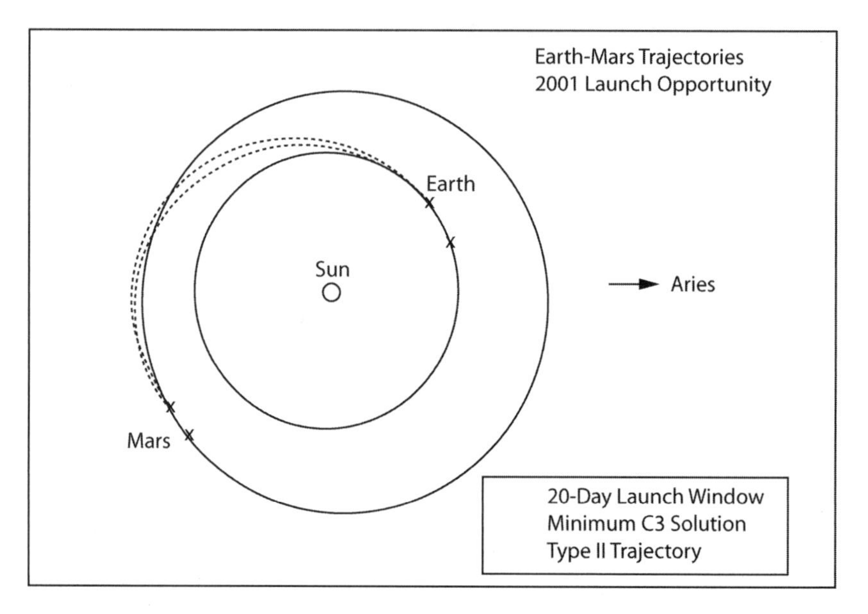

Figure 1. A top-down view of the path a spacecraft must travel from Earth to Mars. Launch opportunities occur every 26 months, and the launch window is open for only 20 days.

out of the grip of Earth's gravity and transfer to a path, or trajectory, ultimately allowing the spacecraft to catch up with Mars and fall into the pull of its gravity. The launch energy requirements outside of this 26-month opportunity are generally well beyond the ability of any rocket we have on Earth. There are some special cases where a mission might go at another time—for example, while the typical duration of a trip to Mars is 7 to 9 months, there are other trajectories that will transport the spacecraft from Earth to Mars in 2 to 3 years—but these launch opportunities have not been used because of the additional time required and the resulting cost to the program.

If you want to send a spaceship from Earth to Mars, this means, in effect, that you have to take advantage of this repeating launch window to reach the red planet. This window is only about 20 days long. Because of the short window, missions have occasionally been sent in pairs, spaced days or a week apart rather than waiting the full 26 months that would be required for a whole new opportunity. This was one such occasion. A month after MCO disappeared, the far more challenging mission, the Mars Polar Lander (MPL), having been launched shortly after its companion craft, also completed the journey to Mars and was set to enter, descend, and land on the surface of the inhospitable Martian south pole. Signals were received as the space probe approached. Then, stunningly, as with MCO, all transmissions suddenly stopped. Once again, scenarios were put forward and calculations were made, and again they were in vain. MPL had disappeared as well.

Two such failures, one on top of the other, were unthinkable to NASA. We were good at space. We had been to Mars recently with the very successful Mars Pathfinder mission. We knew how to do this. The public was equally dismayed. The visibility of the Mars failures and the importance of the Mars program to NASA were such that a review board of high-profile experts was commissioned to find the causes of the failures of the two missions. The results of that board's findings ultimately sent me, a senior civil servant from the NASA Ames Research Center in Mountain View, California, to Washington, D.C., to completely restructure NASA's approach to Mars exploration and to create the first true Mars Exploration Program.

Why do we go to Mars and why does it matter so much when missions fail?

Mars has fascinated humanity for thousands of years. The Roman Mars and Greek Ares gods of war were named for that fuzzy red spot in the sky. Mars was said to have fathered Romulus and Remus, the founders of Rome, and was widely worshipped. Astrologers have forecast the future affairs of men and women based on where Mars was in the sky at a birth or other critical moment in the tide of history.

Mars, the red planet, visible with the naked eye, namesake of March (once the beginning of the calendar year), has inspired poets and writers from Homer to Ray Bradbury. The symbol for Mars, an abstract shield and spear, is the very symbol of manhood. Mars was drawn by Galileo and was once thought to be the home of an advanced civilization. An Italian astronomer, Giovanni Schiaparelli, in 1877 observed lines across Mars through his telescope and called them "canali," literally translated as channels. The popular press and public interpreted canali as canals—the work of some intelligent creature. It is no accident that the popular literature of the late nineteenth and early twentieth centuries as written by Edgar Rice Burroughs had flourishing Martian civilizations. The romance of distant Mars was such that those civilizations came complete with scantily clad women and well-muscled, Tarzan-like men in loincloths. In 1898 H. G. Wells made Mars the home of those coolly calculating intelligences that invaded Earth in his novel *The War of the Worlds*. One of the "canalists," Percival Lowell, an American astronomer and mathematician who predicted the existence of Pluto, built his own observatory in Flagstaff, Arizona, expressly to watch what he believed was the growing of various crops on the Martian surface.

In more recent times, astronomers began to theorize about the nature of the formation of our solar system and surveyed our neighboring planets as a way to imagine other, more distant worlds. Researchers speculated about the possibly organic nature of the dark areas at Mars's poles because the expanses could be seen growing and then receding as the Martian seasons came and went. Those basic surveys established Mars's orbit, size, seasons, and likely atmospheric chemistry at a crude level. However, good solid information about the true nature of Mars was limited to remote observations by Earth-based telescopes until the space age.

Just saying the name "Mars" invokes images, hopes, and dreams for most of us. One critical reason is that Mars is tantalizingly within reach of humans and is surely a target for eventual human exploration. Modern science, particularly the cross-disciplinary field of astrobiology, has shown us that life exists in virtually every niche on Earth that has liquid water at least some part of the year. True, we have walked on the Moon, but this orb, despite recent confirmation of water ice at both lunar poles, is not likely to harbor any living organisms, nor is it a truly compelling site for a second human outpost. Mars, on the other hand, could very well support life of some nature, recognizable to earthlings, and it remains our best hope for a second home in the solar system.

Humans have been adventurers from the days when our ancestors boldly walked out of Africa; we are driven to explore. It is in our nature, and seems often to bring out the best in us. We have accomplished space travel. We have been to the Moon. We want to go to Mars, no matter how hard it might be. And we want to succeed.

After several months of work, the Mars review board concluded that the most probable causes of the double mission failures were two mistakes that were compounded by the largely junior and inexperienced makeup of the teams involved, coupled with inflexible demands that had been imposed on the two projects. In the case of MCO, a confusion of metric and English measurements spelled failure. For MPL, a pennywise and very pound foolish elimination of a full end-to-end systems test meant that a vital line of software code was left out, spelling doom. It is a truism of accident investigation that there is rarely a single flaw that creates a major disaster. Inevitably, a complex chain of events and decisions results in a problem, any one of which, if properly fixed, can prevent tragedy. Mars missions are no exception.

The history of modern Mars exploration is a cautionary tale. Mars is hard. At the time of the MCO/MPL failures, thirty-four missions had been sent to the red planet, beginning with the earliest attempts by the Soviet Union in 1960. Of those first thirty-four missions, only twelve achieved any measure of success, even the sound of that critical first signal from a spacecraft that says it has arrived. Of those twelve, only six could be called fully successful, providing a solid measure of data return. There are good reasons, of course, why Mars is so hard. Launching anything out of Earth's gravity to another world means that each additional pound of mass requires added fuel to reach the escape velocity speed of about 42,000 kilometers per hour (25,000 miles per hour) that breaks gravity's connection to the craft's hardware, and each added allotment of fuel means more mass to launch in a never-ending cycle. The result is that every possible ounce of weight must be carefully considered and pared down to the barest minimum. There is no room for waste. Volume is an equally constraining fact of life in the space business. Propulsion systems, avionics systems, communications systems, data-handling systems, science instruments, and every last detail of what is needed to remotely operate what amounts to a science laboratory that will be millions of miles away demand that every element of the mission be thought through and tested and retested before launch. Once the rocket is fired, the team never again lays hands on the spacecraft. No last-minute adjustments can be made. No hardware can be fixed or replaced from Earth, although new software can be transmitted to fix certain types of problems. Any effort where the overall full success rate is less than 20 percent must be approached carefully—especially considering the hundreds of millions of dollars invested. The odds are against you.

The Soviet Union (and, later, Russia) has been singularly unlucky in exploring Mars. The space program that in 1957 produced Sputnik, the first Earth-orbiting satellite, and in 1961 launched Yuri Gagarin as the first human in space has attempted twenty missions to the red planet. Of that number, only four were minimally successful, and none could be called fully successful.

The United States has a much better track record. The 1964–1971 Mariner missions, while suffering a few setbacks, showed that we could get to Mars, could successfully get a spacecraft into Mars orbit, and should be ready to try for a lander. The Viking mission, which was operational on the surface of Mars from 1976 to 1982, arguably remains a high-water mark in the exploration of Mars; recently, though, much smaller missions have vastly eclipsed the data return. Both Viking 1 and Viking 2 were ambitious combination projects with an orbiter and lander for each. The orbiters took reconnaissance data and provided relay capability for the landers to enable higher data rates for communications. Most of us who were around in the 1970s remember the thrill of seeing the first color pictures of a distant alien world. Nothing was rushed or given short shrift on Viking. The science and engineering teams were the best the nation had to offer. In today's dollars, the total mission cost has been estimated at over $4.5 billion, a flagship-class (largest) robotic mission by anyone's standard.

Those dollars for Viking were used to develop a dazzling array of new technologies: supersonic parachutes, soft-landing thrusters, radioisotope power for operating around the clock, color cameras, lightweight heat shield materials, and a biological instrument package that was miniaturized to create an earthly laboratory the size of a shoebox.

Tens of thousands of images, both orbital and landed, were returned. A definitive analysis of the atmosphere led to the discovery of meteorites from Mars found in Antarctica and identified by their trapped gases. Carbon dioxide frost and identification of water ice in the polar caps all contributed to a revised view of Mars.

In 2000, the MGS mission scientists reported streaks on cliffs that appeared and disappeared rapidly, possibly suggesting water flow on Mars now. Pictures of much higher resolution from MRO show many such "runoff channels," providing stronger evidence that water does flow on the red planet today when ice melts beneath the surface in the summer.

The Viking Program represents a stunning success for NASA, but even the well-funded Viking missions were not without their tribulations. Each Viking spacecraft was really two, an orbiter and a lander, combined and launched as a single craft that separated upon arrival and capture into a Mars orbit. The orbiters successfully lasted through both their primary and extended missions. NASA always defines a primary mission as one that is judged by "full success and minimal success" criteria, which are put in place before launch. An extended mission, which everyone hopes for, is often considered in the initial planning but, given the demands of the space business, may or may not be achieved.

The Viking 1 lander settled into the Chryse Planitia or "Golden Plain" of Mars, a site chosen primarily for its engineering qualifications. For technical reasons it's much easier to land a spacecraft near the equator of a rotating celestial body, and a nice flat plain is always a safer spot to approach than anything that might be cratered or excessively rocky. Although the Viking orbiters took many pictures in an effort to verify the safety of the landing site, their fundamental camera resolution was limited to something on the order of 10 meters (30 feet) in the very best case. As a consequence, the Viking 1 orbiter did not see "Big Joe," a rock 3 meters (10 feet) in diameter that lay only 7.6 meters (25 feet) away. Jim Martin, the legendary Viking project manager, told me many times that he was lucky—the rock was big enough to flip over the lander if the lander had come down on top of it.

The lander eventually lasted through both its primary and extended missions, but not without mishap. Mars really is hard, and landers are extremely hard. The lander's seismometer, one of more than twelve science and engineering instruments, failed to deploy correctly. A locking pin on a sampler arm was stuck and took five days to fully deploy. Still, such minor anomalies taken in context seem irrelevant. A complex science laboratory had been blasted off to another planet, landed without cracking up on a never-before-seen alien landscape, and more or less worked according to plan—a truly stunning success. The lander operated for roughly six years until November 11, 1982, when a faulty command was sent by the ground control crew. The command was intended to uplink new battery-charging software to improve the lander's deteriorating battery capacity, but it inadvertently overwrote data used by the antenna-pointing software. All attempts to contact the lander during the next four months, based on the best educated calculations as to where the antenna might be pointing, failed.

A more significant failure, or, at best, ambiguity, is that of Viking's biology experiment, which produced no definitive results. Some members of the science community were so eager to find life on Mars and so convinced that it was there that they designed experiments to directly detect life without considering that that might not be as easy as it sounds. The experiments that flew with Viking were designed to culture microbes, as might be done in any biology lab, and detect the signs of their metabolic activity using samples from the top few centimeters of Martian soil. The experiments failed to produce evidence of extraterrestrial life or even find any organic (carbon) compounds. Scientists later determined that chemistry and radiation at the surface of Mars made that environment hostile to life as we know it, and so these experiments were essentially doomed from the start. In addition, biologists now know that fewer than 1 percent of microbes on Earth can be cultured in the laboratory. As a consequence of this perceived scientific failure, NASA sidelined so-called exobiology experiments for space missions, especially to Mars, for about twenty years. Recent data reported on the perchlorate (a type of chlorine compound) content of Mars soil suggest an explanation of the failure of the Viking mission to detect organics, but the jury is still out on that conclusion.

NASA is a far more politically driven agency than most of the public realize. We think of the thrill of space, the heroism of the Astronaut Corps, and the phenomenal technological achievements of the agency on behalf of the country. Few of us think about battles for budgets, competition for ever more limited research dollars, decisions made to preserve jobs rather than to advance science and exploration, and the host of crossing winds that batter NASA. Another thing few people seem to realize is how small the NASA budget really has become. When NASA had the frontline mission to demonstrate America's technological prowess during the Cold War of the 1950s and 1960s, NASA received almost 4 percent of the federal budget. As soon as the Apollo Program had done its job, however, and put us at the top of the heap over our then enemy, the Soviet Union, NASA ceased to be such a priority.

In Richard Nixon's administration, the budget slipped downward precipitously and by the mid-1970s had settled at just under 1 percent of the federal budget. Since then, it has often not even been adjusted for inflation and so has become closer to one-half of 1 percent. Some may argue that, because the overall federal budget has been growing, NASA has a total amount of money that is as large as ever. This is not accurate. The overall NASA budget can be looked at in two ways, in constant dollars or as a percentage of the overall federal budget. NASA's budget was the highest it has ever been during the Apollo Program, reaching $33.514 billion in constant 2007 dollars in 1965. For the 2009 fiscal year, NASA's budget was $17.2 billion or 0.55 percent of the overall federal budget. For the 2010 fiscal year, NASA received about $18.7 billion, or 0.52 percent of the federal budget.

Looked at one way, $18.7 billion seems absolutely stingy for an agency that conducts research and development in aeronautics and space technology; operates a number of robotic space probes, including the Mars rovers and the Cassini spacecraft now orbiting Saturn; operates the space shuttle; builds and operates the International Space Station; and manages several other programs, including educational outreach. Since the years of the Apollo Program, however, NASA's budget has never exceeded 1 percent of the federal budget. NASA's budget has been between 0.5 percent and 1 percent of the federal budget since 1975, though it has risen, with some ups and downs, since that year in constant dollars.

NASA does tend to capture headlines, though. In fact, the round blue NASA "meatball" logo (as it is affectionately known) vies only with Coca-Cola and a few other icons for international branding recognition. The U.S. Congress likes to put NASA on the radar screen; unfortunately, this truth has some consequences that may not always be in service to science and technology. At first glance, the budget woes of the space shuttle and the International Space Station might not seem directly relevant to the history of the Mars Program or specific Mars failures, but, in fact, they are just that. With an audacity that may have overlooked the truth that all of space exploration is hard, as a nation we decided that we wanted and could build a reusable fleet of space-bound vehicles that could launch regularly and carry huge payloads into space—big enough to build an entire space station, permanently orbiting Earth. To garner the substantial funding that these enterprises demanded, and to keep these projects going as their budgets inevitably inflated—space really is hard—everything else suffered from the pressures of the "zero sum game" mentality. That is, with enormous shuttle and space station cost overruns pulling money out of every corner of the agency, science and exploration budgets suffered, and pressure grew to accomplish more with less.

Following Viking, then, the next Mars mission, Mars Observer (MO), was not launched until 1992—seventeen years after the landing of Viking—and was subject to great pressure to control costs. Mars Observer represented an ultimately unsuccessful experiment under such stringent cost reduction. The concept was to use a previously developed communications satellite spacecraft body or "bus" designed for Earth and "simply" integrate Mars science instruments. Unfortunately, the huge differences between orbiting Earth and traveling to Mars drove the cost of the modification to a level comparable with building a new spacecraft from scratch. In the end, MO disappeared on August 21, 1993, just before making its final approach to orbit. The board reviewing the failure found it very likely that a small amount of highly volatile propellant had leaked into the metal propulsion lines and exploded when the engines were pressurized just before the final maneuver.

The administrator of NASA at the time of the Mars Observer failure was Dan Goldin, at times brilliant, at times abusive, and always mercurial. Goldin was a self-described cold warrior who had earned his space management stripes working on the Strategic Defense Initiative (Star Wars) and spy satellite projects for a variety of agencies such as the National Reconnaissance Office. When Dan took office in April 1992, he spent most of his first years battling Congress over the future of the International Space Station. Only later did he begin to demand in his particularly aggressive way a new approach to space science missions he called "faster, better, cheaper."

Goldin's philosophy, along with an emerging small mission program called Discovery, led to the decision to break up the payload of the very large MO into three projects. These missions would then be launched over several window opportunities to travel from Earth to Mars to recover the lost science.

First up in the recovery from the MO disaster would be Mars Global Surveyor (MGS), a spacecraft launched in November 1996 that contained two critical instruments rebuilt from the MO payload: a very high-resolution camera operating in the visible range (the Mars Orbiter Camera or MOC) and a special type of camera that could take pictures in the infrared part of the spectrum. Other instruments were aboard, but the MOC in particular, which operated successfully for ten years, dramatically altered our view of Mars. With a footprint on Mars that was a factor of 10 better than the best Viking pictures, the MOC captured a Martian surface never before seen.

Our world has now been treated to images of shifting sand deserts, ice-sculpted dunes, and soil scarred by dust devils. One set of pictures of the same crater wall taken four years apart clearly shows the appearance of a bright streak that could be water ice, frozen just after it gushed from an underground reservoir. NASA's Mars Reconnaissance Orbiter has recently detected dark, finger-like features that appear and extend down some Martian slopes during late spring through summer, fade in winter, and return during the next spring. Repeated observations have tracked the seasonal changes in these recurring features on several steep slopes in the middle latitudes of Mars's southern hemisphere. The best explanation for these observations so far is the repeated flow of briny water on the Martian surface. Water on Mars may well not be a 2-billion-year-old relic, but a presence today.

Launched in the same window opportunity as MGS but on a separate rocket was the highly successful and publicly engaging Mars Pathfinder mission with its airbag landing and toaster-oven-sized rover. Pathfinder would eventually play a significant role in the new Mars Exploration Program.

The next mission to recapture lost MO experiments was the Mars Climate Orbiter (MCO), launched in 1998. As we now know, MCO and its companion in 1998, the Mars Polar Lander (MPL), were doomed almost from the start. The projects were put into a box of tight schedules, constrained budgets, and unchangeable requirements from which they could not escape. The "glue" that holds missions together, systems engineering, was squeezed out of the mix with disastrous consequences.

Dan Goldin had been thrilled with the wildly successful 1997 Mars Pathfinder mission. The public had become engaged with and eagerly followed the adventures of the seemingly intrepid, highly anthropomorphized, little rover *Sojourner*, named after civil rights crusader Sojourner Truth following a national contest. Goldin decided that he wanted more of the same, but with budget pressures created by the increasingly costly Space Shuttle Program and the over-budget International Space Station, he didn't want to pay for it. He dictated that he wanted two missions for the previous price of Pathfinder, and he wanted them in a hurry. He had also promised the White House that he could "do more with less." He pushed for what became his "faster, better, cheaper" approach to space.

Earth, third rock from the Sun, orbits that star in what we define as a year, a bit over 365 days. Mars, the next planet out, orbits the Sun in roughly 687 Earth days or 1.88 Earth years. The conjunction of these two paths results, as we have seen, in a launch window roughly every 26 months during which it is possible to launch a mission from Earth and have it arrive at Mars along the shortest possible path, usually taking somewhere between seven and nine months to get there, depending on mission size, fuel capacity, and other engineering considerations. Goldin wanted two full missions in the 1998 launch window, an orbiter (the type of mission that typically performs reconnaissance tasks) and a lander (the type of mission that provides "ground truth" and very detailed, if geographically limited, data). NASA was to send one of each, during the same window.

Both the MCO and MPL missions were managed from Pasadena by NASA's Jet Propulsion Laboratory (JPL). The contractor project teams that actually built the probes were from Lockheed Martin in Denver. This sort of arrangement is not particularly unusual at NASA where there is a long history of working with aerospace contractors and at times there is political pressure to outsource tasks to private industry. During the development of MCO and MPL, the JPL and Lockheed Martin project teams were constantly faced with the rigid nature of the schedule, budget, and objectives imposed from on high.

The coupled missions, both slated to take advantage of the next biannual Mars launch opportunity, saved money by eliminating the usual layers of checking and double-checking that are common in the "one strike and you're out," unforgiving space exploration business. For the MCO, one of those layers eliminated was the second pair of eyes on the oh-so-critical interface documents. These are the papers that set out the ways in which the flight hardware (the spacecraft with its various systems and science instruments) and software work with each other. It is here, typically, that the conversion of units is checked. The U.S. aerospace world has been dealing with matching metric hardware and U.S. (English-unit) hardware for fifty years, so, no big deal—right? Except, in this case, a measure of rocket thrust was incorrectly translated. A second review by an experienced person would likely have caught the problem. Compounding the error was the junior navigator on duty as the spacecraft approached orbit. The young navigator had the feeling that the radio signal and altitude numbers that were coming back were not quite right, but he didn't have enough experience and confidence to raise the warning flag. The review board determined that the spacecraft entered the atmosphere of Mars perhaps 60 kilometers (36 miles) too low and then burned up.

A similar set of problems doomed MPL. Again, financial constraints led the project team to compromise standard procedure and take more risk by eliminating testing steps. It is customary to conduct a final check of the spacecraft hardware working with the flight software. This was not done for MPL. Rather, a piecewise checking of various elements of hardware and software was carried out and then the entire system was declared ready by "analysis." The flaw in this approach was that a line of missing code went undetected. The review board determined that most likely the lander was in its final descent when the legs were extended and the on-board computer concluded, wrongly, that the spacecraft was on the surface. So, in the absence of any other information, like the radar altitude measurement, the computer said to itself, "Well, the legs are out, I must be on the ground, so let's turn off the jets," and the lander proceeded to crash into the Martian surface at 100 kilometers per hour (about 60 miles per hour). MPL is now likely scattered in hundreds of pieces on that faraway world.

CHAPTER TWO

Getting Started

During the period from 1995 to 1999 when MCO and MPL were being constructed and launched to Mars, I was engaged in two projects, managing the Lunar Prospector mission and then establishing the NASA Astrobiology Institute, and had arguably laid the groundwork for a third (Mars Pathfinder) that set the stage for my eventual role in restructuring the Mars Program, which may help explain why Dan Goldin turned to me, far across the Rockies, rather than a NASA Headquarters insider down the hall.

I had come to NASA in the late 1980s from a small start-up that I had co-founded, nurtured, and eventually sold. I felt it was time to have a little stability in my life and took a job at the Ames Research Center in Mountain View, California. At that time many people, even in the adjacent Silicon Valley, didn't know that there even was a NASA center in Northern California, let alone one that had been there for decades. By the time Goldin summoned me to Washington, D.C., I had progressed to being associate center director, responsible for all the space projects at Ames.

Many projects at NASA—in fact, I would argue many of the best—have longer histories than might be readily apparent to the public. Some of this is because projects that later grow into missions often change names, sometimes more than once, as they develop and make their way from concept to proposal to potential mission and finally to fully funded mission status. Mars Pathfinder began as the first element of the MESUR concept, an acronym drawn from Mars Environmental Survey, a mission proposal for many small landers that I developed at NASA Ames from late 1989 through 1991.

Pathfinder was to prove the technology for the later series of low-cost landers designed to gather data from all over Mars. What it needed to fit into the exceptionally tight cost box that constrained it was a robust entry, descent, and landing (EDL) system. Viking had required parachutes and retro rockets and quite sophisticated landing controls, all of which cost money in themselves, plus they added "mass," or extra weight, which means a bigger landing package, and this really begins to drive up the costs. It's all part of the "Mars is hard" scenario. Pathfinder needed to demonstrate a low-cost way to land that could be repeated in unknown and potentially rugged terrain just in case there were more rocks like Big Joe around. The MESUR study, eventually to be Pathfinder, began in late 1989 when automobile airbags were just becoming standard on some high-end European cars. I started thinking about airbags and wondering if they could protect a small spacecraft from a crash landing the way they were protecting drivers from auto crashes.

My proposal to NASA Headquarters in April 1990 was a low-cost approach that used the venerable, but smaller and cheaper than usual, Delta II rocket for the first time in a planetary mission, employing a disposable "cruise stage" to provide the navigation and propulsion adjustments that would guide the probe to Mars. Unlike the Viking mission in 1976, this spacecraft did not orbit Mars first, but rather went nonstop from Earth through the thin atmosphere of Mars to the surface. Coupled with the airbag EDL system, the mission concept was estimated to cost only about $180 million in 1990—far less than previous landed missions. After about a year of study, the mission concept was adopted by NASA Headquarters, assigned to JPL for development, and eventually became known as Mars Pathfinder, the 1997 mission that carried *Sojourner*, the little rover that could.

The early 1990s also saw the definition of what became known as the Discovery Program, which would further exemplify Dan Goldin's "faster, better, cheaper" approach to missions. The program would fly well-defined, efficient, highly focused science missions, cheaply and often. The idea was to avoid the "Christmas-tree approach" wherein mission opportunities were so relatively hard to come by that every science group around wanted to hang their own ornament—that is, science instrument—on any available spacecraft, knowing that another opportunity might not come for many years or even decades. This approach had plagued NASA from the agency's inception and was notorious for causing schedule delays and cost overruns.

Unlike so-called strategic missions that were assigned to specific NASA centers, the Discovery missions would be competitively selected, a process that was open to the entire science community and that also forced early definition and decision making. Cost caps would be put into place to ensure that missions stayed in a well-defined box. Because Ames had come to be excluded from any new strategic space mission assignments, I eagerly embraced this new initiative as a way to keep Ames in the space mission game. Teams formed rapidly, usually involving a NASA center, a science team, and an industrial partner. At Ames we organized several teams for the competition, and I put significant attention on one in particular—a collaboration between Ames and our next-door neighbor, Lockheed, in Sunnyvale. In the proposal to NASA Headquarters we named the concept Lunar Prospector (LP), and to my delight, we were notified in March 1995 that the concept had been selected for development.

LP was to be a poster child for the new Discovery Program. From late 1995 to May 1998, I was the NASA manager of the LP mission, which, at $63 million, was the lowest-cost planetary mission ever attempted and the first competitively selected Discovery mission. The price tag was, at the time, less than what Hollywood put into a typical action movie. Our "prime" aerospace contractor was the newly merged Lockheed Martin (formed from Lockheed and Martin Marietta) in Sunnyvale, with NASA oversight for the development. Unlike many previous NASA contracts, we had functional requirements—things like "get to the Moon"—and purposefully left the specifics to the contractor. This is similar to the way most of us would build a dream house—figure out our needs and then hire a good general contractor to put it all together. But in this case, the house may be millions of miles away, and it was a new way of doing business for NASA.

Like MCO/MPL, my team faced short schedules, fixed science requirements, and a highly constrained budget. We were squarely in the crosshairs of the faster, better, cheaper era. Although Goldin never really defined what that phrase meant, he clearly indicated that we government types should just write the checks and stay out of the way of the contractor. Trying to use some common sense in this unforgiving business, I found ways to involve our team when I thought the mission was at risk; we also provided both insight and oversight when they were needed. At one point, a critical instrument, the gamma-ray spectrometer, needed some help, in my opinion. This happened to be an area where I had some personal expertise as well as a group of Ames engineers on which to draw. We stepped in and managed to significantly improve the likely functioning of the spectrometer. In the end, it was the right thing to do.

LP arrived at the Moon in early 1998, on schedule and on budget. Over the next eighteen months a wealth of new data was returned, the most exciting of which was strong evidence for water ice in the permanently shadowed craters at the poles of the Moon. This ice could be both a resource for future human exploration and a cosmic deep freeze containing the record of water and carbon delivered to the Moon and Earth over the last 2 billion years.

Another NASA Ames activity that paved the way for how I would approach Mars exploration was the development of astrobiology. In the early 1960s, NASA had created working groups, divisions, and programs around what was called "exobiology" to evaluate the possibility of life on other planets, but the study was dominated more by geologists and comparative planetologists than card-carrying biologists. In fact, exobiology was a bit of a backwater. Then, in the mid-1990s, a perfect storm of science discoveries laid the groundwork for what was to become a new approach to the timeless questions of "Are we alone?," "How did we get here?," and "Where are we going?"

In 1994 the Human Genome Project had met its five-year goal one year ahead of schedule, and its scientists were making great strides in understanding DNA. The Hubble Space Telescope was announcing a continual stream of space science firsts, identifying and photographing objects ranging from black holes to brown dwarfs. In December 1995, a truly startling discovery was announced—the first confirmed detection of a Sun-like extrasolar planet, Pegasi 51. Then, in August 1996, David McKay published what became a very controversial article about a Martian meteorite in the journal *Science* at almost the same time that the orbiter from NASA's Galileo mission began returning intriguing images of the Jovian moon Europa. The Martian meteorite, found in Allan Hills, Antarctica, in 1984 and labeled Allan Hills 84001 (ALH84001), showed evidence of something that could be interpreted as microscopic fossils of primitive, bacteria-like organisms. The images of Europa held compelling evidence of ice rafts floating on what might be a liquid ocean. Tantalizing revelations across a broad spectrum of scientific endeavor were appearing that lacked only definition and direction to bring them into focus. The establishment of astrobiology as a well-defined multidisciplinary science with a true emphasis on biology on Earth, both historic and at its early limits, backed by an institute to fund the new points of emphasis and new approaches to old issues, provided the new path.

NASA Ames held the initial workshops and spearheaded the community, organizing events to bring together scientists from dozens of different disciplines and define astrobiology, the study of life in the universe. Just one result of those efforts was the creation of the NASA Astrobiology Institute, which I was directed to establish in 1998; it is still going strong to this day with more than six hundred senior researchers conducting research through the institute.

By the time MCO and MPL were about to arrive at Mars in September and December of 1999, LP had met its prime mission goals, and I was well into my new astrobiology assignment for Goldin. After being rebuffed by the White House and California politicians in his attempt to close, or at least radically downsize, NASA Ames in 1995, Goldin had needed a new role for a NASA center, and astrobiology, along with information systems—Ames is, after all, in the heart of Silicon Valley—had filled the bill. The centerpiece, the Astrobiology Institute, would be groundbreaking. It would be a true virtual institute where research groups worked across many boundaries of traditional science and institutions. Goldin asked me to be the founder of this organization and make it work. He further told me that I had to find a "King Kong biologist" who could take on the job of director after I had things going. In an amazing bit of serendipity, a Nobel Prize winner named Baruch (Barry) Blumberg was visiting at Stanford University. In the end, he agreed to be the director of the NASA Astrobiology Institute. Mars, of course, is one of the prime targets in the solar system for a detailed study of life in the universe—astrobiology.

During the period from the loss of MCO and MPL until March 2000, the distinguished panel of experts labored behind closed doors to examine this pair of failures. As the due date for the report approached, one of the key recommendations began to leak out: The Mars Program was deeply flawed in part because there was no single person at NASA Headquarters in charge.

In early March 2000, I was in Big Sky, Montana, investigating an earmark. Montana's U.S. senator, Conrad Burns, had put language in NASA's funding legislation calling for the establishment of an astrobiology research project to be established near Yellowstone National Park, and I got tapped to check it out. Was this good science or fluff? A conference was called to show off the research to NASA.

During the second day of the meeting, a Thursday, the cell phone of a colleague, Ed Heffernan, rang. Heffernan was Goldin's chief of staff. He gestured to me and held out the phone. "It's the boss," he said.

Goldin was brief and to the point. "I'll be in California, Manhattan Beach, on Saturday," he said. "I need to talk to you about something very important—can you be there?"

"Yes, of course," I responded. Refusing a direct request from Goldin was considered a career-limiting move.

"Okay, put Heffernan back on," Goldin directed.

Ed nodded and made some notes, then hung up. "Here's the plan," Ed explained. "Meet Dan and me at the Manhattan Beach Hyatt Hotel at 2 p.m. on Saturday. He'll explain it all then."

"Ah . . . could you give me just a little hint on what this is all about?" I inquired tentatively.

Ed fixed me with a quizzical stare. "You know somebody has to go fix the Mars mess, don't you? Do I need to spell it out?"

Forty-eight hours later I was looking at the blue Pacific and listening to the leader of our nation's space program tell me in his unique and colorful way that "your country needs you, son." Goldin asked me to come back to NASA Headquarters and fix the mess that the Mars Program had become—as soon as possible.

Later that night, back home near San Francisco, I had two conflicting emotions: "What a chance," quickly followed by "but Washington is a snake pit" . . . and back again.

In the end, I had that sense that occurs but a few times in the life of a current flowing, of divergent paths coming together to a single moment where one decision can change your life forever. All the things I had done—the Pathfinder concept, Lunar Prospector's success, founding NASA's Astrobiology Institute—had led me to this moment. I had the chance to take a nearly blank sheet of paper and create what would become a decade of missions. How could I not take this chance?

Later on, I found that there were a number of people who thought the Mars Program was not fixable. On Monday morning, I went to NASA Ames to tell my boss that I was accepting Goldin's offer. On my way out of the building I ran into one of those doubters.

"I hear you're riding in to save the day—just be sure you don't get any mud on you and your white horse," he said with more than a touch of sarcasm.

I turned and walked away with a feeling that I had glimpsed the future in a way not granted to many. Mud or not, I would create something that would serve the people for a long time to come.

On March 21, 2000, the Mars Program Independent Assessment Team, led by aerospace veteran Tom Young, presented findings on national television as to the complex of issues that had brought down the two Mars missions, Mars Climate Orbiter and Mars Polar Lander.

The technical issues that had directly destroyed the two spacecraft were seen as the result of deeper systemic flaws that made failure likely, if not inevitable. The principal failure according to the investigative team was that the combined MCO/MPL project was put in a rigid box of schedule, requirements, and budget from which there was no escape—except by taking foolish risks. That report recommended a number of fixes to address all the problems. Chief among them was the appointment of someone at NASA Headquarters to be clearly in charge of the entire program. This was the job that Goldin would hand to me. The position was officially labeled "Mars program director," but a senior member of the media created a new nickname in an article. I would forever be known as the "Mars Czar."

In addition to requiring a recognizable, well-defined, designated leader for the Mars effort, a second less well-known issue that required attention was the culture of risk taking and even bravado that had evolved at JPL in the wake of the huge success of the faster, better, cheaper Mars Pathfinder mission. This cultural issue first came home to me at the scheduled critical orbit and landing events of the two doomed missions.

I had experienced some of that JPL bravado firsthand in the MCO control room during the first disaster. That September 1999, Barry Blumberg and I had traveled down to Pasadena and taken our seats in the VVIP section at the "Lab," as everybody calls it. Dan Goldin and senior JPL managers were stationed in a TV-studio-like arrangement so they could see and be seen as the data came in that would indicate the first mission of a pair, the Mars Climate Orbiter, had successfully arrived at Mars.

This kind of event, the insertion of an orbiter around Mars, was not new to JPL. But JPL was not in direct control of the spacecraft. Lockheed Martin, in Colorado, which had built the spacecraft, would send the signals and monitor the progress of getting into orbit. The events would all be seen in parallel at JPL by the managers who had overall responsibility.

The stage was set. T-shirts, hats, and memorabilia of all sorts were on sale in the JPL courtyard. VIPs received gift bags loaded with pictures, keepsakes, and space goodies of all kinds. A young Arizona State University scientist named Laurie Leshin (currently at NASA Headquarters) was drafted as spokesperson to narrate the events on live television. We had merely to wait for the signal of success as the spacecraft was nudged into orbit by a final burn of the propulsion liquid carried by MCO. The atmosphere was abuzz with optimism bordering on certainty, despite the objective enormity of the task at hand.

We waited. The clock in the control room was set to reach zero just as the signal should appear, when the craft would reappear from the back side of Mars, happily settled into its new orbit around a planet approximately 200 million kilometers (about 120 million miles) away.

Zero hour came. Goldin and the NASA leadership were all live on NASA TV waiting to cheer.

We all know that the signal never arrived. Blumberg, unquenchably curious, began to ask me what was happening. He had been primed by the festive atmosphere and earlier briefings by JPL staff to expect a slam-dunk success. "Perhaps this is just a little glitch with the Deep Space Network," I ventured. The Deep Space Network (DSN) is the array of antennas spaced around the globe that receive and transmit signals to these far distant missions. Maybe the final bit of propulsion put the spacecraft into a slightly different orbit and that accounted for the time delay. I had gone through a similar orbit insertion with Lunar Prospector about a year before and so knew pretty much what to expect.

Minutes went by with no signal. I offered another thought: "Well, maybe the spacecraft had some upset and went into 'safe mode.'" This is the electronic equivalent of hunkering down in the bunker when unexpected problems appear. The software that monitors the health of the probe might detect a fault created by something like a random cosmic ray hitting the computer. If that happens, the craft will begin a predetermined process of examining itself and either phoning home or waiting for instructions from mother earth. Safe modes can sometimes take hours or even days to resolve.

The live NASA TV began to show a somewhat uneasy Dan Goldin.

NASA almost always shines in the crunch. Some of the best space engineers and bona fide rocket scientists in the world leapt into action. Over the next few hours, the JPL control room began a new kind of buzz with tests, ideas, discussions, and, soon, worries. As the evening wore on, it became clear that this was not some simple, easy-to-fix problem.

Antenna stations all over the world were sent to work searching for a signal, but to no avail.

The live TV shots showed final pictures of worried and upset executives. Those of us who had seen Goldin in action began to wait for his famous temper to explode in a fury of expletives and accusations. Fortunately, the NASA public affairs office had the good sense to cut to the studio and begin to fill in with occasional bulletins. Hanging around the VVIP area (Blumberg and I had gold stars on our badges—sort of an "all-access" emblem), I was pleased and surprised to see that Goldin kept cool.

Blumberg and I repaired to the hotel bar and began to muse over the likely outcome of all this. Was the Mars Program headed for limbo as had happened when the Mars Observer mission disappeared a few days away from orbit in 1993? Would the errant MCO spacecraft make a marvelous and amazing reappearance in a few days? Would anybody else care?

"Look on the bright side," Blumberg said. "There's another mission arriving in a month or so. When it gets to Mars, all will be well."

"Except," I responded, "the next mission is the really hard one. The Mars Polar Lander not only has to get to Mars, it has to land, as well, at the south pole, no less. If NASA was gonna goof up on a Mars project, you'd expect the problems to be on the hard one."

It was no little surprise that the world cared a great deal. "Missing Mars Mission" blared across headlines. TV talking heads offered breathless accounts of how NASA may have blown it. The pressure was really on now.

Over the next few weeks, every scenario imaginable, with the exception of alien abduction, a thought espoused only by a few external space enthusiasts, was exercised at JPL in a vain attempt to find a signal from MCO. In parallel, the engineering teams that had worked on the MPL began a hurried analysis of its design. Maybe there was some common flaw that could be fixed by software before the lander arrived. But modern spacecraft are very complex beasts. There are more than 10,000 individual parameters or settings that needed to be exactly right for a typical mission to be successful. Without a lucky find or a flash of inspiration there simply wasn't time to wring out all the "what if" possibilities. In addition, MPL had to get safely to the surface, not using the somewhat inelegant but robust Pathfinder airbag technique, but by employing the retro-rocket scheme not demonstrated since the Viking landers of 1976.

December 3, 1999, dawned bright and crisp, another beautiful winter day in Southern California. Once again, Barry Blumberg and I made the forty-five-minute flight from the San Francisco Bay Area to the Burbank airport. Flying over the vast stretch of studio sound stages that resemble warehouses, I thought how easy Hollywood had made space travel look. "Put some actors in front of a green screen, add a flock of twenty-something digital effects programmers from Industrial Light and Magic, and the director can whisk us to Mars or Alpha Centauri without a care."

The reality is that space exploration is still in its infancy and extremely hard to do successfully. Only about 4,500 launches have occurred in the first fifty years of the space program. During the first fifty years of aviation, more than a million aircraft were built and used, most of them multiple times.

At the time of the MPL landing, the party atmosphere was decidedly muted. A strand of tension was stretched tightly through the crowd and most visible among the professionals in the group. We knew the degree of difficulty on this dive had gone from a 5 to a 10 and the Mars Program, as we had known it, was probably on the line. The Lab's vaunted reputation as a planetary science miracle worker was under threat. Careers hung in the balance.

I recalled the time a year before when the NASA senior executive in charge of science missions, Wes Huntress, had pointed to me and said, "You are responsible for the success of the Lunar Prospector mission. You will have the final go/no-go call for NASA when we poll the stations in the countdown." In January 1998 we were using a new Athena II launcher that had never been used before. The single-stage version, an Athena I, had one success and one very spectacular failure to its credit. The first LP launch attempt was delayed by a range radar malfunction. On the second try, I was pretty certain that the launch would go. Looking up at the Moon, I thought to myself, "Well, Hubbard, you will either get to bask in the glow of success or your NASA career will be over in short order." Lunar Prospector had worked, and I still had my job. That evening at JPL, waiting for the touchdown of the Mars Polar Lander, I could imagine the MPL team reviewing their careers.

Once again Blumberg and I took our VVIP seats and watched and waited. We were all painfully aware of the ten minutes it takes for a signal to travel from Mars to Earth. The viewing screens and clocks were all calibrated to account for the time difference. Unlike the MCO, the MPL had to transmit a signal from the surface, which would be much weaker, so not much data would be in that first message. Mars Global Surveyor, successfully in orbit since 1996, could serve as a relay once operations were under way, but given the disaster of a month before, the expectations were set lower. A simple "Hi, I'm here safely" would have been sufficient.

Following the tradition that NASA conducts all its business in the full view of the world, the public affairs people once again placed Goldin and the other senior people in the eye of the camera. Once again the appointed time came and went without a peep from the spacecraft. To say the expression on everyone's face went from anxiety to despair does not do the situation justice. The product of several hundred million dollars and years of work by hundreds of scientists and engineers had apparently just vanished on the surface of another world.

There was another flurry of options discussed, tests made, alternatives examined. The MGS spacecraft was dutifully tasked to both listen and look for any evidence of MPL. National security assets, as America's collection of spy satellites and listening posts is euphemistically called, were asked to see if there was any trace of either mission to be found. But nothing emerged.

The world was shocked and saddened by this pair of failures, especially coming as they did just a few weeks apart. One of my scientist colleagues in Spain told me that he and his friends felt good when a NASA mission succeeded, and now they were distraught, as the Mediterranean temperament dictates. NASA had failed, so somehow the whole world was worse off.

Tom Young and his team of experts identified two technical reasons for the failures. For MCO, the budgetary pressure caused the project leaders to use almost all junior staff and to eliminate the second pair of eyes traditionally used to catch mistakes. In this case, an engineer had made a mistake in conversion from English units to metric. While the world had a good laugh at NASA's expense, the fact is that NASA and its aerospace contractors have been working across such unit conversions for more than fifty years. Managing these interfaces is the usual responsibility of a seasoned systems engineer. That role was missing in MCO. To compound the problem, the navigation engineer on duty when the spacecraft was to be put into orbit was quite inexperienced, and there was no senior backup. Thus, when the time came to fire the thrusters and make the final orbit determination, the junior navigator did not raise an alarm when the numbers looked less than perfect. Most likely the spacecraft entered Mars's gravity 60 kilometers (36 miles) too low and subsequently burned up in the atmosphere.

The origin of the failure of MPL was more subtle, but driven by the same budgetary, schedule, and performance pressures. It is a truism of the unforgiving space business that you must "test as you fly, fly as you test." The expanded version of this is to test exactly the hardware and software you intend to fly under the conditions they will encounter to the best of your ability to create those conditions on Earth and to fly only what you have fully tested. In their rush to finish the MPL and stay on budget, the project staff made the decision to eliminate a full test of the flight software with the flight hardware. Instead, the testing was done in selected pieces and the product was declared finished "by analysis." In this "one strike and you're out" world, a single line of code was missing. That bit of computer instruction would have told the central computer to check the landing sequence against the altimeter, so as to know that actions like cutting off the retro rockets would not occur until the spacecraft was on the surface. When your mission is on the average 225 million kilometers (135 million miles) away, you had better have done all your critical testing before you left. Along with the loss of MPL, two probes called DS-2 (for "deep space") disappeared. Those would have demonstrated a subsurface measurement capability. Tom Young's review found that the DS-2 test and engineering were woefully lacking and thus the probes should not have been launched.

I had my mandate to "fix the mess." It had been about three weeks. I live near San Francisco, where March is typically solidly into spring. The early bulbs had all gone by, and the wisteria in my backyard was beginning to fade. I had had to convince my long-suffering wife to get our rather large house ready to be rented out; put most of our valuables in storage; get ready for an as-yet-undetermined span of time on the East Coast, where it actually snowed in the winter and was unbearably hot in the summer; and generally completely disrupt our lives for possibly several years. We had flown out to D.C. for a quick reconnaissance trip and found an apartment to rent near a Metro line. The decision that should have taken weeks or even months had been made in minutes. At Ames everyone assumed that I would be gone for at least three years. This was a serious mess, and no one had any idea how to fix it or how long that might take. The folks who didn't think I would be back in three years instead thought I would never come back.

From the time I gave Dan Goldin my acceptance of the program director job around March 14, there was a huge flurry of bureaucratic activity in the background. Travel documents and position reassignments had to be fashioned and approved. In addition, I was immediately brought into the discussion of Mars Program status by Goldin's associate administrator for space science, Ed Weiler, and Weiler's deputy, Earle Huckins, which included the decision of what to do about the 2001 missions. The public affairs people also sent me drafts of the press release that was to go out in concert with the Mars Program Independent Assessment Team (Tom Young Committee) report on March 28. That press release was the one that officially named me as the first Mars program director. Of course, being Washington, all this effort was conducted in the background so as not to leak the story before it was fully baked. As a consequence, when I arrived for my first official NASA Headquarters business day on April 3, I had already been "on the job" for about three weeks.

With a backdrop of a pair of internationally visible failures, the spotlight of the press, and the scrutiny of an impatient administrator, I had already begun to think about what the scope of the "fix" might be. To me what was required was no less than completely reconstructing a queue of missions that would serve at least a decade of scientific goals, and I didn't want it to be a fool's errand. My guess was that we would be putting forward about $10 billion worth of missions for the period 2000–2010, and we needed to get them right.

The problems were epidemic: technical, programmatic, cultural, political, and budgetary. As the Tom Young report stated, leadership at the top was the first order of business. My first day at NASA Headquarters on E Street SW in Washington, D.C., I found that there were at least five people who were certain that they were in charge of the Mars Program. This lack of clarity, responsibility, and accountability is a common failure in bureaucracies where finger pointing and backroom maneuvering are a regular political sport. I would have to find a way to let all five of them know that, in fact, I was now in charge.

NASA Headquarters is located at 300 E Street SW in Washington, D.C. "HQ," as it is now routinely called by all who use it, has been in this building now for perhaps fifteen years. Before locating to this new facility, NASA HQ was in an old federal office building. I remember visiting FOB 10, as they then called it—Federal Office Building 10—in the late 1980s. That typical old federal building, dating probably from the 1940s or so, had huge corridors, fraying rugs, and wonderful tall ceilings with windows that opened for real air. Around the building were artwork and interesting old desk furnishings from who knew where.

The new building, built and then leased to the government, was constructed in an area in Southwest Washington that at the time was in very poor shape. Washington is divided into four quadrants. The fanciest quadrant is the Northwest. The quadrant with the most crime is the Northeast. The Southeast and Southwest areas, which border the Potomac, had fallen into decay. Over the last twenty years a whole series of new office buildings, rental properties, and town homes have been built in the SE and SW, so that these two quadrants are now in relatively good shape.

However, this building at 300 E Street SW, between 3rd Street and 4th Street in the Southwest quadrant, seems to have been built to some awfully minimal standards. The corridors are narrow. The air circulation is poor. For most of the last fifteen years, no wall decorations were allowed. There are two very long corridors on either side of the building with offices branching off. Each floor is dedicated to a different endeavor. The ninth floor houses the administrator's suite and is the seat of power. The third floor is the science floor, home to Ed Weiler, the head of the entire science group at HQ, and therefore home to the Mars Exploration Program. The sixth floor is the aeronautics floor. The seventh floor is space operations, and so on. When I first arrived at NASA HQ, my immediate thought was that it looked like two very long bowling lanes. This was the environment in which all the major decisions for the nation's space program—theoretically the most thrilling, cutting edge, and forward thinking of our federal agencies—were made. The lack of decoration made the building feel sterile and lifeless, and overall the building had a very depressing feel to it.

I was fortunate to be given an office that was relatively spacious and had windows that looked out on the busy freeway next door. I was, however, separated from the rest of the team I eventually assembled as my NASA HQ staff. One near-term goal that was eventually accomplished was to get the whole Mars Program staff together in one spot so that you could facilitate discussions, rather than have to overcome barriers and make decisions more quickly.

At the time that the building lease was negotiated the people at NASA who worked on the details were driven to an extraordinarily low standard of occupancy. If you were a junior staff member or a secretary you got less than 75 square feet to call your own, usually in some type of cubicle where the air circulation was poor. Except for the administrator, who had an entire suite that included a personal conference room, I think the largest offices were perhaps 250 square feet and were used by a tier of officials just below the administrator, called associate administrators. Most fairly senior staff had maybe 120 square feet.

Another facet of the building at 300 E Street SW was its nearly incomprehensible labeling and organization of offices. The best way to describe this is that it's a combination of alphabet soup and numerology. Of these two long corridors, one was numbered from 1 to 50 and the other was numbered from 50 to 100; however, one set of offices was also labeled A through R, and the other set was labeled S through Z. It took you a while to figure out that one corridor had the low numbers, one corridor had the high numbers, one corridor had the first part of the alphabet, and the other corridor had the rest of it. So when someone told you they were in RQ 46 you should allow, if you were visiting there for the first time, at least an additional fifteen minutes because you would invariably go to the wrong part of the wrong corridor on the wrong side of the building. It was impossible to believe that this labyrinthine structure didn't impact people's ability to communicate and coordinate. Although the work was challenging and the charter from the administrator and Ed Weiler, my new direct boss, to create a new Mars Program was one of the most interesting jobs I've ever had, I can say that the building itself, in which we had to work every day, feels like a soulless rabbit warren that impacts virtually everyone and everything in it.

One of the flaws in the Mars Program that, in the view of the Tom Young investigative committee, led to the loss of two important missions was the lack of clear management responsibility and accountability. This manifested itself in very diffuse responsibility at NASA HQ.

There was a solar system exploration division headed by a very good guy named Bill Piotrowski. Ken Ledbetter led the engineers and program executives (a program executive is the HQ engineer-manager responsible for monitoring and reporting the progress of a mission). The science of Mars planning was led by a good friend, Carl Pilcher, who is now the head of the NASA Astrobiology Institute. There were also program executives and program scientists for the Mars missions that were under way in early 2000. Finally, of course, there were Ed Weiler, the associate administrator for space science, and his deputy, Earle Huckins.

So it was that when I arrived, my observation, something that had been validated by the Tom Young review, was that you could find at least five people at NASA HQ who all firmly believed they were in charge of the Mars Program. This was not the fault of any of these people, all of whom were dedicated government employees and good engineers and scientists. However, over time, various pieces of the Mars Program—individual missions, engineering requirements, science requirements, funding profiles—had become the responsibilities of at least these five people, if not more. Consequently, when JPL managers or contractors at places like Lockheed Martin didn't get the answer they expected from, let us say, Bill Piotrowski, they would go and talk the issue over with, say, Carl Pilcher. Or if the engineering management of the program executives wasn't going in the direction someone at JPL thought it should, they might take the issue up with Earle Huckins or they might talk to Bill Piotrowski again.

To address the problem of lack of clear management responsibility and accountability identified by Young's committee, the team recommended that there be a single point of contact at NASA HQ for the oversight, management control, and review of the Mars Program. This new position, the one that I held, was known as the Mars program director. One of my very first tasks when I arrived in D.C. in April 2000 was to work with all the other individuals who had current responsibility and be as clear as possible about my new role.

Fortunately I already knew, had respect for, and had a good working relationship with Piotrowski, Pilcher, Ledbetter, and of course Huckins and Weiler. Weiler and Goldin both helped me a great deal by making it eminently clear that the Mars program director job had the single point responsibility for developing a new program, executing it, and overseeing the mission and technology work already under way.

In D.C., as it is with most government and even industry responsibilities, disagreements and arguments about turf or authority can be reduced to the golden rule—he who has the gold makes the rules. So it was that one of the first things done to underscore the responsibility I had as Mars program director was to give me budgetary authority—scooping all the bits and pieces into which the Mars Program had been fragmented into a single line and then being certain that I had the responsibility to sign off on the expenditures for that budget.

While I often had Dan Goldin breathing down my neck, the actual structure of NASA has a layer below the administrator called associate administrators who are the people with real day-to-day responsibility for billions of dollars. The associate administrator for the Office of Space Science (OSS), Ed Weiler, had the portfolio that included things like the Hubble Space Telescope—and the Mars Program. This made Ed my closest colleague. I had worked with Ed before on a number of programs and found him to be very much a set of contradictions. Sandy-haired and always tan, Ed loved water skiing. I think his perfect goal in life would be to retire to UC San Diego as a professor of astronomy and surfing. However, the professional Ed was that of a brusque, demanding, critical manager, who also was crafty, canny, and respected by the staff at the White House's Office of Management and Budget (OMB) as well as on Capitol Hill.

Following the recommendations of the Tom Young report, I reported directly to Ed Weiler—no doubt and no middlemen. However, in an organization where the key staff are permanent civil servants who can almost never be fired, much more effort is required to effect a change than in the corporate world.

In the organization present in spring 2000 when I arrived, there was a Mars program scientist, a Mars senior program executive, a number of Mars junior program executives for the various Mars missions under way, a chief of engineering with Mars Program responsibility, a chief scientist with Mars science oversight, and of course Ed Weiler and his deputy, Earle Huckins.

In this thicket of competing agendas, I set to work, using my mandate from the top, diplomacy, logical persuasion, and emotional appeals to organize a new Mars Program staff. In this cat-herding exercise the only thing I really controlled was where I put the cat food. That means the money.

In my three-hour discussion with Dan Goldin about taking the job as Mars Czar, one critical negotiation point was my firm demand to be unequivocally in charge of the Mars budget. Goldin and Weiler made it clear to the rest of the HQ staff that I (a) reported to the top and (b) was delegated responsibility for the entire Mars budget, then about $400 million a year. Once the money flow was clarified, the hearts and minds of the NASA HQ staff followed.

Setting up a brand-new Mars Program Office was much more than finding a few offices to occupy. Federal bureaucracies are rules-driven entities, so a host of new rules about reporting, approvals on funds transfers, performance appraisals, and the like had to be written or modified. I would occasionally pause and reflect on the amazing ability of a government organization to conduct something as challenging as space exploration, given the enormous constraints of the federal process.

Fortunately, I had the support of Ed Weiler and his unflappable deputy, Earle Huckins. Earle was a veteran of many decades of NASA insider maneuvering; he was an excellent program manager with great experience and a wellspring of calm, southern-flavored advice. Through Earle's help I had a nearly free hand to pick supporting project, resource, and administrative staff. However, after years of downsizing driven by Goldin's directive that HQ be a "policy-only" organization, HQ had only one-third the workforce of a decade earlier. The impact of this was that I had to reconstruct and then manage a queue of missions and a budget of almost a half billion dollars per year with three full-time engineers, two part-time engineers, a part-time money manager, some part-time scientists, and a secretary.

In the NASA space enterprise, the execution of all these complex missions cannot be the sole responsibility of a tiny HQ-type organization. NASA depends on its ten field centers to oversee the management of the current $19 billion per year of taxpayers' funds. Each center has its own culture, history, and fiefdoms. For the Mars Program, the center responsible is the Jet Propulsion Lab. One of the other key recommendations from Tom Young's group was that, just as at HQ, at JPL too there must be a Mars program manager with sole responsibility for both meeting the requirements developed at HQ and the success of the program.

I had worked with JPL for decades, both from Ames and in earlier positions.

There were a handful of people I knew and respected at JPL who might fill the job outlined by Young's group, but all of them had current, high-profile, critical positions. How would we pry loose the right person? Here again, I had the help of Earle Huckins, who understood the institutional arrangements at JPL better than I did. Together we settled on Firouz Naderi, who was currently the manager of what was called the Origins Program of space telescopes. This queue of missions had been put together with a strategic view, and the decade-long total cost was in the billions of dollars. Naderi was already a twenty-five-year JPL employee and someone who clearly knew not only the techniques of large-scale program management, but also the JPL talent and infrastructure. He and I had worked together during the last few years since one key astronomy program centered at NASA Ames was part of the overall Origins Program. I knew Naderi to be precise, articulate, demanding, and utterly lacking in political nonsense. While a JPL "lifer," Firouz could also be critical of his own home institution. This honesty would be key in addressing the problems afflicting the Mars Program.

Earle and I had a teleconference with the JPL director at the time, Ed Stone. Earle explained how Dan Goldin wanted the very best talent applied to fixing the Mars Program. It was clear that Goldin had the Mars Program very high up on his personal radar screen. After some resistance, and a bit of jawboning by Earle, Stone relented and agreed to transfer Naderi to the Mars Program. Without Firouz, the restructuring of the Mars Program would not only have been monumentally more difficult, it's unlikely that it would have been nearly the success it seems to have been. His role deserves a bit of detail.

Firouz Naderi is Persian, one of the many highly educated, very cosmopolitan people who escaped Iran as the Shah's regime was falling and the ayatollahs were beginning to take over. I first met Firouz, as I mentioned, when he was leading the Origins Program for JPL. I noted then that he was a very clear strategic thinker who would make his thoughts known in a direct way and was not fearful of taking on Ed Weiler in a technical argument. The Origins Program was a series of missions designed to look for evidence of what Carl Sagan used to call "the pale blue dot," an Earth-like planet with the potential to harbor life. These were astronomical missions peering out into the universe, looking for evidence of another planet that might have the same life-supporting environment as here on Earth.

When I arrived at NASA HQ in 2000, one of the first management actions that was required was establishing a new programmatic connection between NASA HQ and JPL. The Tom Young review stated very clearly that the existing management path between HQ and JPL was completely broken. There was neither clear accountability nor clear responsibility.

In my discussions with Earle Huckins, we very soon focused on the need for a JPL Mars program manager who could think strategically, had experience with flight missions, and had enough clout at JPL to be able to command the respect of the rest of the organization. It wasn't very long, a matter of minutes, before the name Firouz Naderi arose. The problem, of course, was that Firouz had responsibility for the high-profile Origins

Program. The first week or so of my tenure at HQ, Earle and I had the teleconference with Ed Stone. We had already discussed with Ed Weiler the possibility of asking Firouz to join the Mars Program. Weiler recognized that although Origins was an important program, the Mars Program, with the failures and high visibility, put Mars at the very top of the priority list. In our discussion with Ed Stone, he at first resisted making the move because of the importance of the Origins Program, but as we explained the implications of the visibility and the urgency of fixing the Mars Program, Stone relented and agreed that Firouz could be our JPL program manager.

Shortly thereafter Firouz and I began to speak by phone, but it was clear that we needed an extended planning session that could only be accomplished face-to-face. The best place to do that was at JPL. Although I would end up setting requirements at NASA HQ, the missions would be implemented through JPL, and it was important that we discuss all the problems identified by the Tom Young committee.

Within a matter of a week or less I took one of what would be many, many flights to Los Angeles and drove to the JPL campus. Even so simple a thing as a cross-country flight, it turned out, required strategic thinking. The closest airport to JPL is Burbank, but this is a regional airport, and there are no nonstop flights from D.C. The default airport is LAX, Los Angeles. When you fly from D.C. to L.A. you had better organize your flight to arrive in the middle of the day. Picking up a rental car at LAX is just the beginning of your drive across downtown Los Angeles and up to Pasadena, home to JPL. On a good day with no traffic accidents, that drive is less than an hour. If you arrive at rush hour, which in Los Angeles is anywhere from three in the afternoon to seven or eight at night, you can count on an hour and a half to three hours just to get across town and, even then, the unpredictable often occurs. I learned to plan with enormous cushions for the vagaries of L.A. traffic lunacy.

Firouz and I met many times over the year and a half of planning and rolling out the new Mars Exploration Program (MEP). In our initial meeting Firouz did something that I was not expecting and which set the tone for our relationship of the next ten years. Firouz told me that he'd read the Tom Young report, he was aware that the relationship between HQ and JPL was dysfunctional, and he was determined that we would find a way to work together effectively and professionally. That level of directness is all too rare in most federal agency situations. In addition to meeting and discussing how we would restructure the program, he guided me to his house near the ocean, quite a few miles from JPL. He fixed me dinner, a gourmet extravaganza of Mediterranean delicacies. Over wine and good food we began to learn a little bit more about each other. I have to say that I developed immense respect for Firouz's determination to do whatever he could for the Mars Program. His openness and his willingness to acknowledge areas where perhaps JPL had not performed as well as it should have were both a delightful surprise and an absolute necessity for any kind of meaningful progress.

Those first few meetings were critical in establishing the outlines of how we would go about this restructuring. It was clear to me that the community was dispirited and, because of historical processes, didn't feel like they were involved in the planning of the missions. Firouz kept challenging me to state what the hard edges or hard walls of the constraint box were. JPL and the engineering staff were clearly used to operating in a vise and wanted to know the worst right up front. It was out of this discussion of our constraints and the trades that would be necessary that we began to evolve the notion of program systems engineering.

Both Firouz and I had experience in how programs were developed and how the politics of Washington worked. It wasn't very long before both of us were approached by individuals at HQ, at JPL, from the science community, and from other places with ideas and concepts. Along the way some of these people with their own agendas and their own desires to shape the program would begin to quote one or the other of us as saying something. This quickly turned into a "he said, she said" environment where the potential for misunderstandings and even rancor or hostility might emerge.

As a consequence of this inside-the-Beltway, spin doctor, agenda-setting, rumor-mongering activity that went on continuously, Firouz and I adopted a simple agreement. We agreed that if either of us heard someone quoting the other person in a fashion that sounded strange or odd or contrary to the path we were on, we would not react. As soon as was feasible, Firouz and I would speak over the phone, sort through the different versions in this game of Washington telephone, and decide what it was that we were doing or not doing. This seemingly straightforward agreement spared us from more wasted hours of worrying about what the other organization or the other group was up to than can be imagined.

Firouz and I each assembled our respective teams and chartered them to work with each other. The interactions were not always smooth or easy, but we felt that as long as the two of us maintained open, clear, and honest communications we could work through whatever difficulties we encountered.

Here are some examples of how Firouz and I divided the responsibility for restructuring and implementing the program.

As the program director at HQ, I would establish the program policies and objectives. JPL and Firouz would implement the program all the way through the projects and take end-to-end responsibility. I would assess and evaluate the Mars Exploration Program and its projects. Firouz would establish the performance metrics and provide regular reports to me. I had the responsibility to define the Mars exploration strategy and establish the architecture and road map. Firouz supported the development of the strategy and the architecture road map and was responsible for conducting the studies that would identify or flesh out candidate missions for future opportunities.

I established the program-level budgets and budget profile in the top-level allocation. Firouz developed the details of the budget requirements, managed the detailed budget, and did the sub-allocations of the budget. One way we described this was that, given the pressures and hectic pace at HQ, I managed a 20-line budget. At JPL, Firouz had a 200-line budget with a much greater amount of detail. I could look at that detail anytime I wanted, but in the Washington environment what's most important is your knowledge of the strategic directions and what's happening at a top level.

I would select the missions to implement our architecture and road map, subject, of course, to the approval of Ed Weiler and at times the NASA administrator as well. JPL's job was to implement the approved missions, manage activities that were undertaken in other centers like NASA Ames Research Center and NASA Langley Research Center, and implement the technology that would enable future missions. I approved the program plan. JPL and Firouz would develop the program plan. My requirements were called level I requirements. That meant that they were the topmost instructions, directives for everyone else to follow. At JPL in the program office, they established the level II requirements, to the next level of detail. I served as the chief advocate for the program. JPL and Firouz supported me in this and also planned and implemented an education and outreach program based on that advocacy. Finally, it was up to me at HQ to negotiate deals across HQ organizations, across agencies occasionally, and also to lead the negotiations on international agreements. JPL and Firouz were the managers for the interface after the negotiations had been successfully concluded.

This actually all worked out. It wasn't straightforward or obvious to begin with, but after ten years I can say that the commitment everybody brought to the table was substantive, real, and ultimately successful.

In the years preceding my appointment as Mars program director at HQ I had been fortunate to get to know many of the leading members of the Mars science community. These relationships helped me a great deal as I set about restructuring the Mars Program in both my initial discussions with Ed Weiler and after my appointment. It was clear that we needed both a science advisory group and a program scientist at NASA HQ. Formal advice to a federal agency is provided by a duly constituted committee set up under what is known as the Federal Advisory Committee Act rules. Any other group is only able to provide a federal agency like NASA with options, analyses, and the like. So in the beginning of my tenure we set up something called the Mars Ad Hoc Science Team or MAST. This team included people like Ron Greeley from Arizona State University, a former chair of a National Academy of Sciences group; Larry Soderblom from the U.S. Geological Survey; and Jack Farmer from Arizona State University, who was a well-known astrobiologist.

Ed Weiler insisted on one other ingredient—a Mars senior program scientist who would work directly with me. Ed's choice for program scientist was Jim Garvin. I had met Jim briefly before as a result of being the manager of the Lunar Prospector mission. Jim's great strength is in brainstorming the scientific future of Mars exploration. His deep knowledge of past experiments and instrumentation and his awareness of current developments made him an invaluable asset as we worked to determine what the next set of missions would be. It was also clear that Weiler had a great deal of respect for Garvin's scientific insight and his ability to explain the scientific return for different approaches.

Garvin also had one capability that I found truly amazing. He could disappear. Because of this vanishing skill, I nicknamed him the Pooka. A pooka is a mischievous spirit of probably Celtic origin. My first exposure to the term came many years ago watching the famous movie *Harvey*, starring Jimmy Stewart. Harvey is a giant six-foot-three-and-a-half-inch, invisible rabbit seen, at first, only by the Jimmy Stewart character, and then later by a once disbelieving psychiatrist. There is a famous scene in the movie where a dictionary falls open to the word *pooka*. The creature is defined as being here and there, now and then. So it was that I would be talking to Jim Garvin standing in the doorway of his transitional cubicle at NASA HQ about the science of some future mission. I would look away for mere seconds. When I would look back, having not moved, Jim would be gone. The next thing I knew he would be phoning me from NASA's Goddard Space Flight Center in Greenbelt, Maryland, or his home in Maryland, explaining that he had had an emergency task or a special assignment to take care of. I have never been able to figure out how Jim was able to disappear and then reappear miles away. His pooka capability will always remain a mystery.

In late May or early June 2000, a series of photographs from the Mars Global Surveyor (MGS) mission reached NASA HQ. They were startling. They appeared to show what experienced scientists thought were runoff channels, like the result of a flash flood in the desert. The Viking mission had given us a great deal of information and many thousands of images that were often interpreted as evidence of water in ancient Martian times. As MGS continued exploration, coupled with the small amount of data from Mars Pathfinder, the story of ancient water was confirmed again and again. By ancient water most scientists mean perhaps 2 billion years ago. These new pictures, though, showed a Mars surface not heavily pitted with craters, unmistakable signs of age. They showed a Mars surface with what looked like modern runoff channels. This seemed to imply that there was some source of water still beneath the surface that emerged occasionally in what geologists would call the modern era. Now, the modern era to a geologist is not like the modern era to a baseball fan—1900 to date. The modern era to a Mars scientist is somewhere perhaps in the last few hundred thousand years. Nevertheless, to take six zeros off the date of water on the surface of Mars is an extraordinary discovery.

Whenever there is an extraordinary finding from NASA, it is customary to provide this information first to the stakeholders paying the bills. These are the important committees on Capitol Hill and, during my tenure, the office of the science adviser for the president. The science adviser in this case, Leon Feurth, was actually the adviser for Vice President Gore, who was extremely interested in science and the progress in understanding life in the universe.

The NASA public affairs people set up a meeting where Ed Weiler, Jim Garvin, and I visited the Old Executive Office Building (now renamed the Eisenhower Executive Office Building) and spoke to Gore's adviser. We were informed that it would be just Gore's adviser and perhaps one aide. The announcement of modern water on Mars had been submitted to the prestigious journal *Science* and would be released later in the week (June 30, 2000). Our visit to the White House staff was meant to be a courtesy, not the official release of the information, which needed to go through peer-reviewed scientific channels.

We were driven to the meeting in the NASA official car, which is used for high-level executives when they go to Capitol Hill and the White House, and arrived at the Old Executive Office Building. We found our way to the adviser's office. As I remember, Feurth had a terrible cold that day, so our discussion was often interrupted by his sneezing. He was in clear discomfort. Sitting on the sofa, we laid out the pictures on a coffee table. To my surprise there was not just the adviser and his aide, but perhaps six or eight young staffers in the room as well. I learned later on that this is the way of things inside the Beltway. You should never be surprised when far more people show up for a supposedly closed or sensitive or double-secret briefing. Information is power and new and exciting information is that much more power.

We spent perhaps forty-five minutes briefing Gore's adviser. He and his aides asked good questions, and I think we communicated to them how exciting this discovery of modern water on Mars was. I think that Ed, Jim, and I showed that the path we were on could in fact lead to, if not the discovery of life on Mars, perhaps at least a habitat for life.

We left the office, got back in our chauffeured Town Car, and began to head back to NASA HQ. We had not been gone from Feurth's office for more than ten minutes when my cell phone rang. It was the chief Washington correspondent for *Aviation Week* calling to ask me if I had a comment to make about the discovery of modern water on Mars. Clearly what had happened is that one of the staffers in the office phoned a friend at *Aviation Week* to give them the scoop. I learned a valuable lesson from this. Nothing stays secret in Washington for long, and it's best if you cultivate the members of the media and work with them. It happened that the person calling was someone that I'd already spoken with in the past and whom I trusted. Therefore, it was relatively easy to explain that the story would appear through the peer-reviewed journal *Science* and that we would really appreciate it if *Aviation Week* would not put out anything prematurely. Thus it was that we were able to coordinate the leaks and releases so that from the public's perception, we had an organized story about modern water on Mars.

I often thought about this experience with the near instantaneous leaking of information as I proceeded through the next year of planning. As a federal agency we had an obligation to let the public know what our plans were and what we thought the best way ahead was. The danger, of course, is that if incomplete or half-baked or poorly thought-through information is leaked in a haphazard fashion, only confusion will result. The experience with the modern-water leak taught me that it is extremely valuable to clearly think through what your next steps are and then work closely with the media people both at NASA and in the outside press to be sure the story is told clearly and effectively. I would sometimes refer to this as haiku for science. My goal was to express complex scientific or program ideas in as simple and understandable a way as possible.

My initial team, the reins of authority, and the outlines of the immense task ahead were almost set. The final piece of the puzzle was for me to assemble a "kitchen cabinet." Space is a complex business involving an intricate dance of science, engineering, and management. Everything has to be right from the big picture to the tiniest details. There are no second chances in space. I was experienced enough to realize that a group of experts with insights into a broad range of issues would be required to have even a hope of getting this right. I would need all the help I could assemble. NASA had a plethora of committees, both formally mandated by law and ad hoc. I knew many of the scientists and engineers on these groups and thought I would get reasonable advice when the time came. In the meantime, though, I needed something different—a few people who had a combination of experience and vision that could offer me off-the-record, hard-nosed assessments of the program.

Over time my kitchen cabinet came to be about eight of the most experienced Martians I could find, but the two who were the founding members were Jim Martin and B. Gentry Lee. Both were already legendary in the planetary mission community and as a pair were almost like a father-and-son combination. They figured prominently over the next year as I had to make rapid decisions with billion-dollar impact.

Martin was a bear of a man, perhaps six feet four and 250 pounds or more, and reached fame as the project manager of the Viking mission. Many of the people who now are executives in the aerospace business passed through Jim's tutelage and mentorship. Jim had the ability of the best leaders to immediately penetrate to the heart of your proposal and find the weakest point. His "wirebrushing" of shallow thinkers was well known, to such an extent that some fearful managers would try to exclude him from review committees, lest they be exposed to criticism. Jim had been the key reviewer when I presented the Mars Pathfinder concept in 1990 and was the chair of the independent review committee for the Lunar Prospector mission. He was exactly what I wanted in an adviser.

Jim's partner was my friend and colleague Gentry Lee. Gentry's apprenticeship was on the Viking mission, eventually serving as the lead for all the science operations. His grounding in the fundamentals of systems engineering provided a foundation for the type of critical analysis that the Mars Program would need if we were to be successful. Gentry's skills and talents extended far beyond the planetary project world, though. As a full partner with Carl Sagan, Gentry Lee was instrumental in the production of the *Cosmos* series that premiered on PBS in 1980. Together with Arthur C. Clarke, Gentry co-authored the sequels in the *Rama* series and went on to write several more science fiction novels on his own. I can also say that as a raconteur and dinner companion Gentry has no peer. I wanted not only Gentry's systems analysis, but also his sense of the bigger picture and his extraordinary creativity.

Now I had the basics in place, but a host of very-near-term decisions were clamoring for attention:

Every 26 months Earth and Mars align for that brief window of 20 days where a mission can take advantage of what is known as a "minimum energy trajectory." If you miss the 20-day window, even the most powerful launchers available cannot send any meaningful payload to Mars.

Two spacecraft were being readied for launch in about a year. As virtual clones of MCO and MPL, did they also contain fatal flaws? Should we proceed with one mission, both, or none?

In just about 38 months the 2003 launch opportunity to go to Mars would then open up. Should we send an orbiter, a lander, or play it safe and skip the opportunity?

In addition, the French space agency (Centre National d'Études Spatiales or CNES) and NASA/JPL were pursuing intensive planning for a Mars sample-return mission that was to be launched in 2005. The mission was to be "faster, better, cheaper." Did this make any sense? Were the science and technology ready?

It was just the beginning.

CHAPTER THREE

Inside the Beltway

In an ideal world, I would have sat down in my "Mars Czar" office at NASA HQ on E Street SW and started methodically planning how to approach the enormous task before me—get ten years of demanding interplanetary missions planned within the available budgets and make certain they all were successful, that the program was coherent and made sense, and that it would be sustainable to the end of the decade. Oh, and by the way, I had to deal with enormous pressure to integrate a possible Mars sample-return mission into the queue (list of missions). In fact, the issue of Mars sample return became much bigger than I could have anticipated. In reality, no such leisurely approach was even remotely possible.

Immediate decisions loomed like wolves howling at the door. There were two missions under development for launch in May 2001—an orbiter and a lander. I had to understand what shape they were in and quick, and my team and I almost certainly would have to come up with some way to decide whether or not both or either of them should launch. This is the kind of decision that brings center directors, CEOs, and congresspeople breathing down your neck. Jobs, reputations, and millions of dollars are on the line. Inexorable pressures of planets in motion meant that inefficient use of a week or two could easily result in missing the launch window down the road and mean another two years and the attendant very expensive marching army.

It was April 2000. There was barely a year before launch. In the space world this is usually the time when development challenges are behind you. All the designs have been finalized for months. All the hardware is built. Most, if not all, of the software is written. The project focus moves to the final integration and testing of the payload with the spacecraft to ensure that everything works together as it should. Although it is not uncommon to find problems during testing, most difficulties have traditionally been failures of parts (for which there were often already-built "spares" available for just such contingencies) but are now more commonly software glitches that require rewriting and debugging. It is rare, if ever, to challenge the whole development philosophy and process so late in the game. Nevertheless, that is where my HQ team, the JPL Mars Program staff, and the contractors found ourselves.

A bit of a political digression is in order here. The truth is that politics is never truly a digression in the federal world. Everything is politics every minute of every day. One of the key people on the political side of things for me was a fellow I came to truly respect, Steve Isakowitz at the Office of Management and Budget (OMB). Steve is now the chief financial officer for the Department of Energy, but at the time in the spring of 2000 when I was restructuring the Mars Program he was a budget examiner and branch chief at OMB. His job was to review, critique, and recommend budgets for the science part of the Department of Energy, for all of NASA, and, I believe, for the National Science Foundation as well.

Isakowitz and his team were responsible for looking at the planning, the budgets, the technical effort, and the work that went into creating a new Mars Program. I learned that Steve was not a simple budgeteer but rather was an engineer trained at MIT who had worked for a while in the aerospace program at Martin Marietta. This meant that Steve came to the job with an understanding of the engineering and science challenges and an appreciation of them, as well as being very knowledgeable about the federal budget process.

Steve also had the ability to think and probe critically at any assertion you made and any of the technical approaches that you were taking. At first his questions seemed, frankly, a bit hostile, but as I got to know Steve I understood that he was simply trying to be sure that he understood, and you too understood, what was being proposed.

At my first meeting with Steve I was accompanied by Ed Weiler, who introduced me and proceeded to lay out the initial part of the discussion for the restructuring. Eventually Ed got to where he had observed that Steve and I had a good working relationship and that I could hold my own in the questioning. Then, he let me go repeatedly on my own.

A satirical look at the trials and tribulations of the NASA HQ program director with all the gates, traps, and obstacles to be faced. (Courtesy Roger Arno)

During the development of the 2003 mission at JPL, a model of the 25-pound Pathfinder rover (left) engaged in a chase around the clean-room floor with the 384-pound MER rover. (Courtesy NASA/JPL-Caltech)

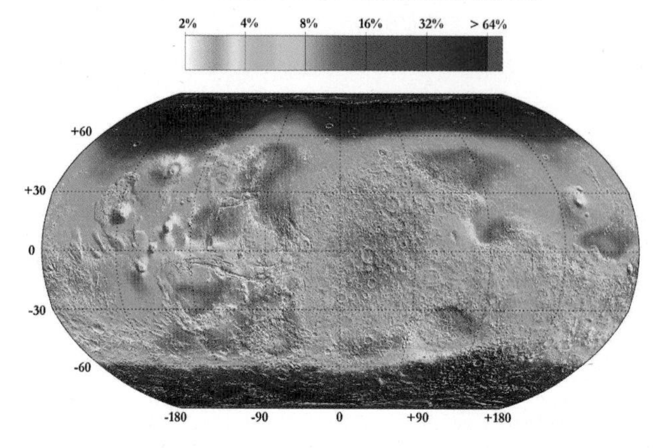

Odyssey's detection of very large amounts of water ice in the top meter (3 feet) of the soil of Mars was a major accomplishment in "following the water." This finding set the stage for the *Phoenix* landing site. (Courtesy NASA/JPL-Caltech/University of Arizona)

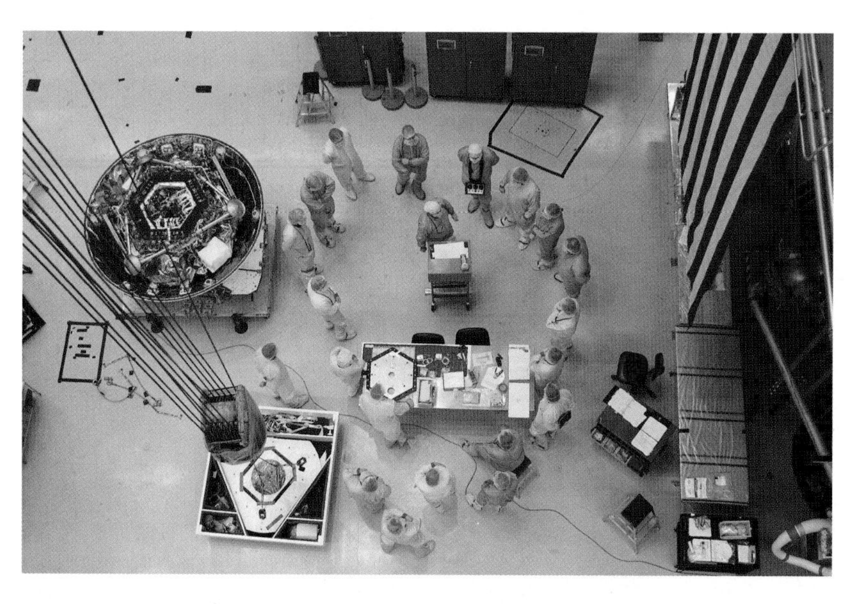

The *Phoenix* lander in final test in the clean room at Lockheed Martin, Denver (2006). Compare the size of this vehicle with the much larger Mars Science Laboratory rover (last plate). (Courtesy NASA/JPL/University of Arizona/Lockheed Martin)

MRO captured an avalanche of fine-grained ice and dust, possibly including large blocks of water ice, in process on Mars in February 2008. This was an amazing example of how dynamic the red planet is today. (Courtesy NASA/JPL-Caltech/University of Arizona)

This MRO image shows part of Gale Crater, the site chosen for the Mars Science Laboratory rover because of the 5-kilometer (3-mile) mountain of water-related minerals in the center, right next to a safe landing area—identified by the 20-kilometer (12-mile) ellipse. (Courtesy NASA/JPL/University of Arizona)

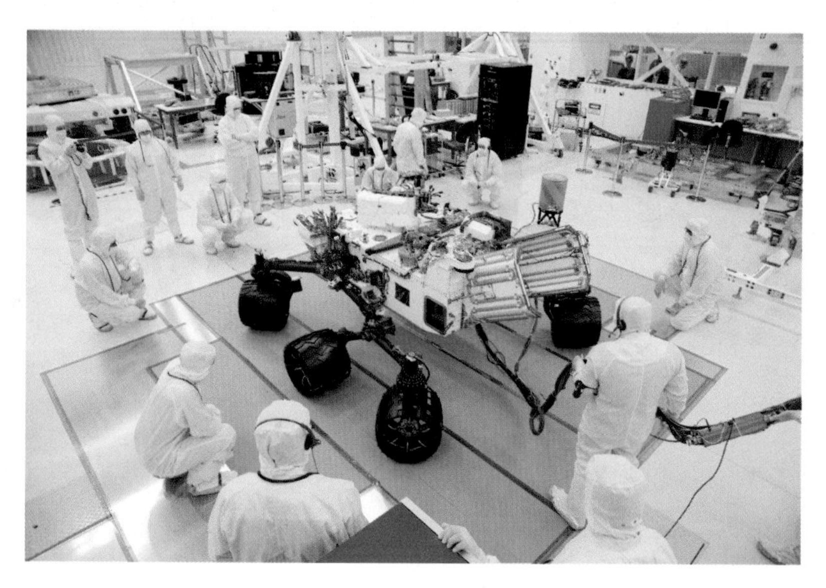

The Mars Science Laboratory rover in the JPL clean room during testing. At nearly 1 metric ton (2,200 pounds), this vehicle will be the most sophisticated and capable robotic lab ever placed on another world. (Courtesy NASA/JPL-Caltech)

The important part of the federal budget cycle is the interaction between the agency and the budget examiner, such as Steve, beginning in the spring of any given year. This time is called the spring preview, and it gives a heads-up of what the agency may be thinking about the budget plans for the following year. This sort of planning usually takes place around May or April. As the summer progresses, there are periodic meetings between key people in an agency and OMB. The critical date is usually around Labor Day. That's when the agency formally submits the budget to OMB for its consideration for the following year. The next critical milestone is called the passback. The passback usually occurs around Thanksgiving, and this is the reaction, the formal reaction, of OMB to the agency's proposals.

Following the passback, during the holiday season is when the so-called reclama process takes place. The reclama process is where the agency responds to OMB's passback, which may contain cuts, may contain increases, and may contain language that can be either prohibitive or supportive. In chapter 9, I'll describe the reclama process that took place during the holiday season of 2000 that resulted in more than $500 million being added to the Mars Program budget.

Following the reclama, OMB makes the final decisions, assembles all this into the president's five-year budget, and prepares that package to be released after the State of the Union speech. It's important to note that although Congress enacts the budget one year at a time, the OMB and the White House prepare five-year budgets. Ed Weiler always told me that if you got your new program or your new budget into the five-year statement by OMB, that was as good as it got. Congress has a staff of about 20,000 people; there are 2 million civil servants working for the executive branch. As a consequence, what Congress can do in any given budget year is usually a major up or down vote. There'll be times when specific items approved by OMB are voted on, and, of course, the dreaded earmarks, but for the most part, at least in any reasonable political environment, having a program endorsed by OMB usually means that it has a high likelihood of getting approved on Capitol Hill.

While mentally juggling multiyear budgets in my head, I still had to deal with the immediate decisions to be made about the 2001 launch opportunity and whether to fly either or both of the two spacecraft already being developed for Mars. Managed by JPL, but being constructed at Lockheed Martin in Denver, the two missions were holdovers from the previous faster, better, cheaper approach that sought to send both a lander and an orbiter to Mars every 26 months on a bargain basement budget. As found by the Tom Young investigative review, the defining characteristics of this style of mission were very lean management, tight budget, minimal checks and balances, and great faith that, left to its own, industry could produce a successful space mission at lower cost and more rapidly than a government laboratory. The extreme version of this management experiment was the assumption that NASA staff should just write the checks and stay out of the way of the contractor. I had already seen, as mentioned in chapter 2, that this was not always wise.

Over a period of a few months, it became apparent to me that the 2001 lander and orbiter suffered from many of the same flaws that caused the failure of their 1998 siblings, the Mars Climate Orbiter (MCO) and Mars Polar Lander (MPL).

This approach—carrying a nearly identical lander and orbiter at each launch—was the result of a top-down directive from Dan Goldin. It's a scary business taking on your boss or your boss's boss.

Every spacecraft, whether it's an orbiter or a lander, has a body or "bus" that is the mechanical foundation for the craft and a lot of engineering elements that allow it to maneuver, collect, and transmit data and perform its functions, whatever they may be. The heart of the craft, though, is in the collection or "suite" of science instruments. They form the raison d'être for a given craft and also make every scientific spacecraft unique. To understand a spacecraft, one has to have at least a rough idea of what its instrumentation does and how it does it. A good example is to be found in the lander and orbiter on which I had to pass judgment in a hurry that spring.

The orbiter was designed to carry near duplicates of several instruments that had originally been aboard the ill-fated Mars Observer, lost in 1993. One of the most important of those experiments was a set of sensors that could detect the elements in the top meter (3 feet) of the Martian soil from 400 kilometers (250 miles) above the red planet. Called the GRS (gamma-ray spectrometer) instrument suite, the unit included the GRS, to detect gamma-ray photons, plus a neutron spectrometer and a high-energy neutron detector to detect neutrons released from the surface of the planet. The technique was novel and had first been proposed in the mid-1970s. All the planets are continuously bathed in the subatomic particles of the solar wind, the spray of energy blasting continuously from the sun, mostly positively charged protons. In addition, the rest of our galaxy of stars produces very high-energy particles and radiation called galactic cosmic rays. This constant stream of radiation penetrates the thin atmosphere of Mars and strikes the soil, where it is absorbed.

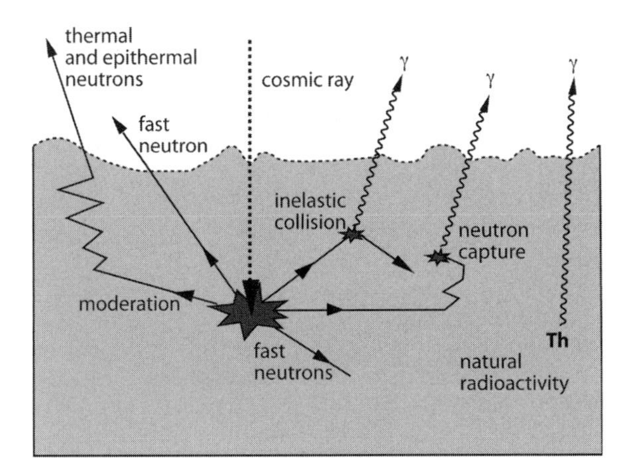

Figure 2. This schematic illustrates how a cosmic ray interacts with the soil of Mars to create nuclear radiation that serves as the "fingerprint" identification of many elements, including hydrogen in the form of water ice. There is always some naturally radioactive material in the soil; here, thorium (Th) is indicated. (Courtesy NASA/JPL-Caltech/University of Arizona)

As each molecule of the Martian surface is struck by the solar and galactic radiation, the atoms of the soil briefly become more energetic, then release that energy back into space in the form of light (photons) and neutral particles (neutrons). The energy of those photons (called gamma-rays) and neutrons provides a unique fingerprint identifying the element in the soil that they initially strike. This technique works best on the lighter elements, so with the importance of water to life, one of the signals being sought was that of hydrogen, the H in H_2O. The essentials of these interactions are illustrated in figure 2.

This instrument succeeded beyond the wildest dreams of the scientists when it was eventually carried to Mars aboard the 2001 orbiter, *Odyssey*. A hydrogen map was constructed using the results. The map assumes that hydrogen is essentially equivalent to water, H_2O, the likeliest case. This kind of map is a very common way of presenting science data and is an extremely useful tool for the science community. Even a lay audience, though, can see that at the poles of Mars, the water-ice concentration is greater than 60 percent by mass fraction (more than 80 percent by volume).

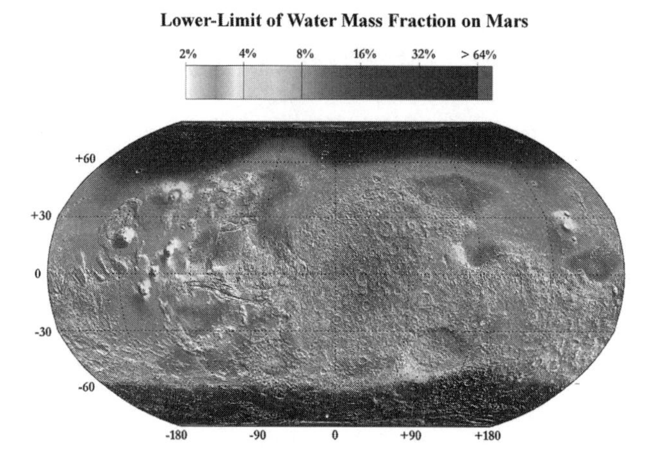

Figure 3. *Odyssey*'s detection of very large amounts of water ice in the top meter (3 feet) of the soil of Mars was a major accomplishment in "following the water." This finding set the stage for the *Phoenix* landing site. (Courtesy NASA/JPL-Caltech/University of Arizona)

Given the strength of the signal and the freezing temperatures at the surface of Mars, the only possible interpretation was that an enormous quantity of water ice was present in the top meter (3 feet) of the Martian soil.

I was familiar with all of this from research early in my career at Lawrence Berkeley National Laboratory and my time as the NASA manager of the 1998 Lunar Prospector (LP). We had both a gamma-ray spectrometer and a neutron spectrometer aboard LP and similarly were pleased and surprised at the amount of excess hydrogen, which could only be understood as water ice. The proposed 2001 orbiter, then, seemed well designed, well understood, and likely to give us a good science result. The case was by no means as clear for the Martian lander.

Shortly after I arrived at NASA HQ, my local staff and the new Mars Program Office at JPL held a rapid review of the orbiter by a team of very senior managers and engineers who found over a hundred "must-fix" items, but no mortal flaw. By contrast, the 2001 landed mission incorporated an essentially untried landing system that had to work in order to even begin to give any science instruments a chance to function. Sadly, we had neither the time nor the budget to adequately test the proposed poor man's version of the retro-rocket propulsion system that was last used in 1976 by Viking. As the Mars Program Independent Assessment Team had detailed in the

Tom Young report, the Mars lander flight software was never tested with the flight hardware, leading to the loss of the mission. What made us think it would be better this time around?

Lessons learned about the MCO failure and the MPL failure were being applied in evaluation of the Mars 2001 projects. When I arrived, there were still some who thought that perhaps both the lander and the orbiter might be launched. Dick Spehalski, who was the former project manager for the Cassini mission to Saturn, was put in charge of the "red team review." Within the aerospace industry, it is common for review committees to have a color associated with them. Sometimes they're the black hat review that seeks to find the weak points in your program. Sometimes it's a blue team or a red team review. The idea is to have people who are not associated with the project but are very knowledgeable to come in and see if you're ready to fly or if you are proceeding on a good development path. Red is often assigned to a critical review team that will make an important decision toward the end of a given process.

The red team looking at the Mars 2001 missions was populated by the Jet Propulsion Laboratory (JPL); the Aerospace Corporation, which does reviews mostly on behalf of the U.S. Air Force; and personnel from Langley Research Center, where there was an outpost of engineers who supported NASA HQ. As a result of these red team reviews, typically very rigorous, there were several major issues raised and those were all addressed in one way or another by the time of the "pre-ship review" in December 2000—the review that assesses whether or not the system or subsystem is ready to go to the launch site or other venue. One major issue that was discovered was that there was no check valve installed in the propulsion line that would provide the energy to put the spacecraft into a circular orbit around Mars. The lack of this check valve was considered to be a major flaw because a previous failure review team had identified this check valve, or its lack, as being the probable reason for the loss of the Mars Observer mission in 1993.

It was pretty clear to the red team that, like the failures of MCO and MPL, the faster, better, cheaper mantra had resulted in engineers becoming less than rigorous about paying attention to the details. Like I always say (and may have first heard Tom Young say), spaceflight is a "one strike and you're out" business. I know that when we eventually launched the Mars orbiter *Odyssey* in April 2001, we had counted over 10,000 different items or different quantities that had to be ensured or constructed just perfectly or else the mission would have been a failure. After the red team review for the orbiter and the attendant resolution, either through design changes or testing or, in some cases, using "as is" (meaning that the risk of using as is had been identified as low), *Odyssey* was permitted to continue toward launch. The key instruments in the Mars 2001 orbiter were the gamma-ray spectrometer, which had been intended to fly on the Mars Observer mission years before, and an instrument called THEMIS, a thermal emission imaging system. Thermal emission imaging means that the instrument is sensitive to the waves that are associated with heat. The THEMIS has allowed scientists to be able to detect the presence of minerals on the surface of Mars associated with water by looking at their emissions in the thermal infrared region. Such complementarity of instruments is common on planetary missions. Data from one instrument help corroborate data from another. The orbiter looked good and the whole team was pretty positive about it. That part of the decision, at least, seemed resolved.

When a mission runs into trouble, one thing you can count on for immediately subsequent, related missions is plenty of reviews. The 2001 orbiter not only received a review immediately after the report from Tom Young was issued, but continued to have reviews before the hardware was shipped to Cape Canaveral Air Force Station (CCAFS) in Florida, as well as other reviews, to be sure that all the findings from the failures had been addressed. Almost as soon as I arrived at NASA HQ, I recall going with Earle Huckins, Ed Weiler's deputy, to look at the status of the 2001 lander. It had always been clear to me that the lander posed far more risk to the success of the mission than the orbiter. NASA has been doing orbiters, along with its contractors in the Jet Propulsion Laboratory, for almost fifty years. Although getting to Mars is never easy and the failure rate has been high over the years, the elements of putting an orbiter around another planetary body are much better understood than the entry systems, the entry conditions, the winds, and the surface of another planet.

When our team went to Denver to look at the 2001 lander hardware, a near clone of the failed MPL, we were skeptical. The presenters were highly enthusiastic and optimistic. They pointed to things they had done to ensure success, like outriggers on the landing legs to be certain that the lander would not tip over. They pointed to the testing that they would now be conducting to ensure that the hardware and software all worked together. But after a few days of presentations Earle and I came to the same conclusion that another red team had reached, which was that the risk posed by the 2001 lander, the deficiencies in its test program, and the number of open issues led to a higher than acceptable probability that the project would fail. I felt great about keeping the orbiter on track, but I realized that I needed to cancel the lander.

Figure 4. After the successful launch of *Odyssey*, Ed Weiler presents Hubbard with an award.

The 2001 lander team was not happy. The Athena science payload investigator, Steve Squyres, was not happy. The Athena payload, an important suite of scientific instruments, consisted of a Pancam (high-resolution panoramic camera) and Mini-TES (miniature thermal emission spectrometer) to survey the scene around the rover and look for the most interesting rocks and soils. Three more instruments—a microscopic imager (MI) and two spectrometers (MIMOS and APXS)—could be placed against these rock and soil targets to gather more detailed data. The rock abrasion tool (aka the RAT) could then be used to scrape away the outer layers of a rock to see what lay beneath. This payload eventually flew on both Mars Exploration Rovers.

Lockheed Martin, the company working on this together with JPL, was not too pleased either. Clearly, though, my job was not a popularity contest. I had to do what I knew to be right. Ed Weiler, who was new on the job as associate administrator when MCO and MPL disappeared, was firmly supportive of this decision. As Ed said repeatedly, he didn't want "two more potholes on Mars."

Years later, I was told by the people who worked on what became the *Phoenix* mission, which took the 2001 lander hardware, installed a new science payload, and ultimately flew it successfully to the surface of Mars, that my decision had been the correct one. Once they started fully testing all the 2001 lander hardware, many more defects and flaws came to light, any one of which would have been sufficient to cause the mission to fail.

There are a lot of risk reduction actions that were taken to ensure that the orbiter, Mars *Odyssey,* named after Arthur C. Clarke's *2001: A Space Odyssey* novel, would be a success. There was an independent verification of all the mission-critical parameters and separate verification of English and metric units. One of the standing jokes about the loss of MCO and MPL was that the Americans had forgotten how to convert English units to metric units. Of course, American industry had been working in both English and metric units for over a hundred years. The problem in this case was that "faster, better, cheaper" had resulted in a lack of personnel reserves, which in turn resulted in lack of oversight to catch any one of a number of mortal mistakes.

Some of the other things that were done to ensure that Mars *Odyssey* in 2001 was successful were the augmented staffing, improved tracking and ranging, adding the validation of the software, and adding the check valves that could have been the cause of the Mars Observer loss years earlier, as well as a host of other details. And finally there was a cultural shift to be certain that the loose organizational connections were firmed up and lines of responsibility, accountability, and authority were made clear. Also, personnel who had been working in development had to be designated to transition into operations. It is often important that people who know and have built the spacecraft are around when the spacecraft is actually traveling or orbiting to another planet so that someone who observes a problem can have the actual engineer who built it assess the difficulty.

In the end, the scientific results of the 2001 mission more than justified all of the effort that was put into the retesting of the hardware. The promise of the gamma-ray spectrometer, in what would become known as "following the water," was fully justified. What we found was that the surface of Mars is covered by water ice at latitudes greater than 60 degrees in at least the first meter (3 feet) or so (the depth to which the GRS could make measurements) as measured by the gamma-ray and neutron technique described above. The measurements from orbit indicated that at the poles there were places where the water-ice content was more than 60 percent by mass fraction (greater than 80 percent by volume). This translates into literally billions of gallons of water, if the ice were melted, and may well explain some of the mystery of where all the surface water for Mars went.

With a solid and, I was confident, correct decision made on the immediate pressing concern of what to do about the 2001 missions under development, I could finally begin to do the job I had come to Washington to do—fix the mess for the long term by creating a true program for Mars and get beyond the opportunistic, disjointed path we had been following. I also intended to do my best to put some assurances in place to keep things successful and on track for at least the next decade.

I already had the right tools in hand with my team. I was especially pleased with Firouz Naderi and Gentry Lee, both brilliant and both true straight shooters. Jim Garvin, the Pooka, was a genuinely fine scientist, and I knew he would be an asset. Of course, I knew I had to have a good handle on the ground rules and that meant fully understanding the environment in which I would be working. My main source for that was Ed Weiler.

Ed Weiler was, as I've said, my immediate boss during the reconstruction of the Mars Program. Over the years I've known Ed Weiler, he's been the source of more than a little controversy, but in the end I count him as a friend of science and one of my friends. Many people have seen only the brusque and demanding side of Ed. They are unaware of some of the personal pressures that he has faced over the years as a single parent. One of his children has been afflicted by some serious emotional disorders, and although he has mentioned this both in private conversations and a few times in public, he has very rarely alluded to this part of his life. I think that any parent who is attempting to conduct a very high-stress job with a seriously ill child must feel an enormous amount of pressure.

My first introduction to Ed on a personal basis goes very far back to about 1990 when he was in charge of the astrophysics program and I was relatively new at Ames, working with some of the people producing infrared detectors for future astrophysics missions. Ed seemed brusque, even abrasive, but at the same time genuinely interested in the science.

My first major interaction with Ed was when I was asked to set up the Astrobiology Institute. By that time Ed was responsible for the entire science organization at NASA HQ. There was an offsite strategic-planning meeting to which I was invited where Ed was working with his staff and members of the science community to put together a three-year strategic plan. After a long day's meeting there was a group of us gathered at the retreat site hotel lobby having a few drinks and talking about the future. Steve Squyres was there and other scientists from the planetary community. Ed wandered up, having just played, I believe, a game of pool with Steve. Ed does not drink, and, in fact, I don't believe I have ever seen him have a drink during all the time I've known him. Still, he sat down at the table where I was, and we began to talk about astrobiology. It seems that, at the time, we were suffering a great deal trying to maintain support for the scientists and science—sadly, a reality that regularly continues to plague us. I recall pointing out to Ed, though, that his program had at the time a total of something like $5 billion a year, and I asked him quite pointedly, and even a little aggressively, why we couldn't have just $1 million or $2 million for astrobiology at Ames. I got into a bit of a heated discussion over this and in the end we agreed to disagree. Years later when I was the "Mars Czar" struggling to balance my budget over the ten years we were addressing, Ed never let me forget how hard it was to fit in even an extra $1 million when you have commitments far larger than you could possibly address in the program at hand.

Another aspect of Ed, of course, is his love of water-skiing and surfing. I believe that he spent many an hour out in Chesapeake Bay during the summer skiing on those waters, and also never missed an opportunity if possible to go to San Diego where some of the best ocean surfing in the United States occurs. Ed is not fond of international travel—he once described flying to Europe as the awful experience of being stuffed in an aluminum tube for as much as ten or twelve hours.

In the final analysis, the thing I respect most about Ed is his commitment to doing good science and working as hard as he can to navigate the tricky, shark-infested waters of Washington to provide good science to the community. The science organization at NASA HQ was blessed for many years with a sequence of highly talented scientific leaders who understood the political process and how to make solid community-based missions happen. First, in my experience, was Len Fisk, who created the three-year strategic-planning process and on whose watch a number of critical missions got started, including my own proposed Mars mission that eventually was named Mars Pathfinder. The second extraordinary leader for science organization was Wes Huntress. Wes was instrumental in getting the Discovery Program going and in creating a program budget line for Mars even though the implementation of that ended up with the two MCO and MPL failures. Wes was also a key player, along with Charlie Kennel and France Cordova, in helping us in Ames found the science of astrobiology and the Astrobiology Institute.

Ed Weiler was the third in this triumvirate, and he inherited the Mars failures, which together we turned into a decade-long string of successes. Ed was best known for having endured the initial problems with the Hubble Space Telescope and living through that to see it become one of the signature successes of science and NASA. One of the things I learned from Ed Weiler was how to sell a program to the Office of Management and Budget. Ed was a master at creating a scientific case that had solid grounding in the community, yet also had public appeal that was understandable by the taxpayers and the managers in both the executive branch and the legislative branch. Prior to my arrival, Ed had already convinced the administration to fund the Origins Program that Firouz Naderi had been working on. Ed's initial instructions about how to craft a program document and description helped me a great deal in putting just the right touch on the materials that would sell each proposed mission in the new Mars Program.

There has been a period in recent years where, in my opinion, NASA science leadership did not perform at nearly as high a level as Len Fisk, Wes Huntress, and Ed Weiler did, but as I write this Ed is back at the helm, as NASA HQ associate administrator for space science, working to ensure that key science missions endorsed by the community and selected by the community get proper funding and that this success continues into the future.

With Ed guiding me through the labyrinthine corridors of Washington's inside-the-Beltway politics, sometimes referred to as the "logic-free zone," and Firouz Naderi, Jim Garvin, and Gentry Lee keeping the science and systems engineering solid, I knew I had a real shot at making a success of what might have otherwise seemed an impossible task.

CHAPTER FOUR

Follow the Water

When I was named to lead the Mars Program on March 28, 2000, the program was driven to no small degree by the desire to return a sample of Mars soil and rock to Earth as quickly as possible. This focus on returning a Mars sample was driven primarily by news and opinions surrounding the Allan Hills meteorite, ALH84001, found in Antarctica in 1984. There are now more than ninety-five of these so-called SNC (shergottite, nakhlite, chassigny) meteorites. These meteorites are known to come from Mars because the analysis of the trapped gases inside them matches exactly the atmosphere of Mars as was measured by the Viking landers back in the mid-1970s. When the Allan Hills meteorite was ultimately analyzed by a very broad team led from the Johnson Space Center (JSC) in Houston, they made some amazing announcements, including not only that this meteorite from Mars contained organic material (carbon-based compounds), but that the researchers believed they saw evidence of tiny fossils looking very much like wormlike or tubelike structures. The French seemed particularly enamored of these intriguing possibilities and began to think the world needed samples returned directly from Mars to prove that life, if not extant on Mars, had at least been there at some point in the planet's past.

These microscopic fossils, called nano-fossils by some because of their extremely small size, resembled, to a degree, Earth-based organisms. However, the size of these organisms, even assuming that they were larger before they were completely dried out, was so much smaller than Earth-based microorganisms that many were skeptical that they were once living things. I would say that, today, the scientific community in general does not believe these are nano-fossils from Mars, but rather some sort of artifact of the way the sample was prepared. However, the useful thing about this debate was that it focused the biological/astrobiological community on the question of how small could life be. Astrobiology is the science of life in the universe. It seeks to understand the basic principles of living organisms and whether they could have emerged anywhere else in the solar system on another of the "pale blue dots" of which Carl Sagan spoke so eloquently. No one had ever thought to start with the size of an RNA or DNA strand inside a cell wall, add some fluid in the cell, and figure out the size of the organism. Such a scientific discussion turned out to be extremely valuable and has truly advanced astrobiology, even if, in the end, the conclusion was that the size of an actual living organism would need to be far larger than the putative nano-fossils from Mars.

I would say the scientific consensus about the Allan Hills meteorite is that it is definitely from Mars and it may contain organics and tiny magnetic crystals from Mars, but none of the other claims seem to be substantiated by the community. Nevertheless, the JSC team, headed by David McKay, still feels fairly strongly that they have seen evidence of organic material from Mars. Why is this important? Organics, carbon-based compounds, are the very stuff of life, along with liquid water. All of the living organisms of which we are aware require both liquid water and carbon to exist.

The Allan Hills meteorite touched off a focused effort by the NASA science organization to bring a sample back from Mars—that is, a well-selected sample capable of teaching us new and compelling things about Mars. Even if the Allan Hills meteorite were shown to have contained organic material or even true nano-fossils, the drawback is that this meteorite has no planetary context. No one knows exactly where on Mars this meteorite emerged from. Without a context, in space science, as in most disciplines from art and anthropology to zoology, isolated information is of limited value in increasing understanding. Of the possible ninety-five or so Mars meteorites we now have identified on Earth, all we know is that they came from Mars. We don't know from where.

The astrobiology community in particular has been quite vocal over the years in saying that while geology and even atmospheric science can learn a great deal from a rock sample taken almost anywhere on the surface of a planet, if we are to understand the processes of life and whether life could have emerged on Mars, we need a sample, preferably carbon rich, that has been around water at some point in its history, and we need the full context of that sample to understand the likely processes it has undergone. Not just any rock will do. It must be well selected. Ideally, we want a sample taken from an area that was probably in the presence of liquid water for a long period of time, shows organic materials if at all possible, and has associated with it some of the other minerals that are often associated, on Earth at least, with biological processes. In early 2000, we didn't know enough about Mars to be able to definitively point to the planet and say, "Here's where all these things are true, and here's where we should pick our sample."

When I began to hold meetings with the broader Mars science and aerospace community in the spring of 2000, there were the beginnings of a sample-return mission in process. The sample return involved the close collaboration, one might even say interdependence, of the French space agency Centre National d'Études Spatiales (CNES) and NASA. There were many assumptions about the cost of this mission. The claim was that for about U.S. $750 million as our contribution, a project could be launched to Mars that would return a sample perhaps as early as 2004 or at the latest by 2006. As my team and I started looking in earnest at the assumptions and the technological requirements of getting from the year 2000 to a launch in 2003 or 2005 to bring a sample back, it became clear that there were at least four major technological hurdles in addition to the scientific selection quandary.

The first requirement was a type of rocket referred to as a Mars ascent vehicle—the thing that would enable the collected sample to get off the surface of Mars on the first leg of its journey back to Earth. Such a rocket would have to be carried all the way to a Mars lander on the surface, and then after a year or so point to a place in the Martian sky and, with 99 percent reliability, launch a sample up to a waiting orbiter.

The second technology was the entire chain of acquiring a sample on the surface of Mars, sealing it—to the satisfaction of those who feared that a dangerous, contaminating alien strain of something might potentially be included—in some type of robust container, and making it available to send back to Earth. To do this, one must also obey the international agreements of a discipline known as planetary protection. Since we don't know for sure whether or not any samples from Mars would contain any organisms that might be dangerous to Earth, nor do we understand exactly how Earth-based compounds and organisms might contaminate the collection area, elaborate controls must be put in place to prevent either forward or backward contamination. A sample, once it gets back to Earth, has to be handled extremely carefully in much the same way that very dangerous viruses and bacteria, like the ebola virus, are handled, with the added issue that we don't want to be finding Earth-based contamination and thinking it's Martian life.

The third major technology, in 2000 at least, was the ability to capture in orbit around Mars a basketball-sized container with the samples and secure it safely to the waiting orbiter with the Earth return vehicle.

Finally, there is the challenge of returning samples to Earth within a probability of breakup or accidental contamination of less than 1 in 1 million, the level of certainty required by advisory groups. Both the Stardust mission (to gather comet samples; launched in 1999, returned in 2006) and the Genesis mission (to gather particles of solar wind from the Sun; launched in 2001, returned in 2004) eventually showed that simply designed containers were actually amazingly robust and returned samples could be quite well protected with existing technology. The Genesis container crashed into the Utah desert due to parachute failure and the container was breached, but a good part of the data was rescued anyway. Much was learned from the event and that knowledge was successfully applied to Stardust; these lessons could be directly applied to future container design. At the time, in 2000, however, the requirements and the ability to meet them were formidable.

A ground-based Mars sample-handling facility had to be developed as well. The facility would receive the sample, keep it under containment as if it were "the Andromeda strain" (a reference to the popular 1969 Michael Crichton novel and 1971 movie of the same name), and allow tests and examination before release of the samples to the broad scientific community. Early workshops on such a facility indicated that it would cost at least $300 million to design and build to meet the likely levels of containment required at that time.

Why all this fuss about a sample from Mars? Can't you simply go, I am often asked, to the surface of Mars with an outstanding scientific laboratory and look for life? The answer is "No, you can't," and there are a number of reasons for this. As I said earlier, there were logical hurdles to be addressed before a Mars sample-return mission got under way. In my mind in 2000, the largest single hurdle was the scientific question. Where do you get a well-selected sample? What are the fingerprints of life for which we are searching? I didn't want to spend billions of tax dollars to find out that Mars has basaltic rocks on the surface—something we already knew. Once I looked at the sample-return program that was under way in 2000 and had a chance to talk to the broader scientific community, I posed that very question to a room full of sixty-five or seventy planetary scientists: "How do you get a sample? What is worth the $2 billion to $3 billion it would cost to bring it back?" I did not get a crisp answer.

What the science community was saying to me by their lack of specifics included a number of very important things. First of all, there is still not even an agreed-upon definition of life. The closest you can get is perhaps seven or eight attributes of life on which most astrobiologists would agree. These are things like the ability to reproduce, being based on carbon, reaction to the environment, capable of Darwinian evolution, and other such characteristics. So to say that you can go to Mars and select a sample that is conclusively life bearing or even potentially life bearing, using only the instruments you can take there, was very problematic in 2000. Gradually, the scientific community, and astrobiology in particular, adopted the more general approach of looking for habitable environments—places where life could conceivably have emerged. This means the presence for long periods of time of liquid water, the presence of organic materials, and the presence of enough energy, probably in the form of heat, to allow living systems to exist and evolve. In other words, look for a reasonable context, something at which we knew we had a real shot at success, and begin to narrow down the search in that way. I characterized this approach as looking for the fingerprints of life—those attributes and characteristics that we knew to be conducive to life. We didn't have a *Star Trek*–type tricorder that would allow us to say, "It's alive, Jim!" so looking for a suitable environment seemed to me the only rational approach.

I had an excruciating decision to make. Since the technologies that were required for Mars sample return had not received a substantial investment and since the scientific community at that time was clearly unsure about how to acquire the right kind of samples, I actually knew what I had to do. I canceled the existing sample-return project. As we will see, however, the push to have a sample return continued to dominate the discussion as we constructed a new mission queue.

There were extremely unhappy people on both sides of the Atlantic, but in the end, even with ten years of perspective, I am still convinced that it was the right decision. Furthermore, with the time that has passed, I would say much of the rest of the scientific community, if not most, agrees with me. In 2007, the National Research Council commissioned a study on the search for a sample to return from Mars. The results of the study, led by Bruce Jakosky, a well-known astrobiologist from the University of Colorado, said that in the intervening years since the two ill-fated mission launches of 1998, the community had learned a lot, enough to be able to say that once the data from existing missions and the planned Mars Science Laboratory were digested, they could at last pick a place that was a habitable environment and from which samples would probably contain some version of the fingerprints of life. The whole stance is a far cry from the "Let's go for a hole-in-one" days of Viking.

Along with pressure to do a sample return, I inherited a top-down directive to fly an orbiter and a lander at every 26-month opportunity. I had already determined that, at least for the immediate opportunity on the table, that didn't make good sense. As we took our nearly clean sheet of paper and began to redefine what we meant by Mars exploration, we began to move toward a science-driven, systems-level process that sought to understand all of Mars, not focus on the process of spaceflight or sample acquisition first. Today I think we would say the Mars Exploration Program is a science-driven, technology-enabled effort to characterize and understand Mars, including its current environment, climate, and geological history and, importantly, its "first-among-equals" biological potential. Central to the questions to be addressed is "Could life ever arise on Mars?" With all these competing opinions, interests, and groups, I knew I needed an airtight approach, a strategy and a set of arguments that would convince and satisfy, if not everyone, at least most of them.

A final complication was that in our planning, we assumed that all of our scientific and engineering measurements of the nature of Mars would be carried out using robotic assets, but we also needed to make allowance for experiments that would provide critical information for the eventual human exploration of Mars, and we incorporated this planning, too, into our integrated approach. We needed to get it right.

To begin, we had to recognize that we had certain constraints. Firouz Naderi at JPL and I spent quite a bit of time figuring out, first, what were the hard walls of the program box. That is, what constraints were out of our immediate control. While obviously we were given a budget with which to work, although in the end we were able, thanks to the help of the NASA administration and the OMB, to increase our budget profile, in the beginning we had to assume that what we had to work with was pretty well fixed.

Second, nature and physics gave us a 26-month launch opportunity cycle. Every 26 months for a brief 20 days, there is an opportunity for that minimum energy trajectory to Mars. Figure 1 in chapter 1 shows the basic path of how a mission to Mars works. You launch from Earth. You get out of Earth's gravity well, transfer to a different orbit—you're basically catching up with Mars—and then you either go into orbit as an orbiting spacecraft or through a difficult and challenging entry, descent, and landing

(EDL) program and end up on the surface. Our final big-picture constraint was the readiness of both the science and the instrumentation, as well as the technologies to allow us to explore Mars.

As was stated in the description of the Mars sample-return program present in the year 2000, it was my conclusion and that of my team that a sample return was technologically and scientifically unready. We had to evaluate what is known as our "risk posture." The risk management had to be consistent with the visibility of the Mars Program. We couldn't appear to be reckless with such an important program, nor could we be timid. Again, we had to get it just right. Along with the International Space Station, the Mars Program was one of the handful of projects that the administrator himself paid close attention to. Added to that, as we found when MCO and MPL disappeared in late 1999, the entire world judged NASA by the success of its Mars missions.

Our emphasis turned first to systems engineering, that integrated and careful discipline at which both Firouz Naderi and Gentry Lee were such world-class experts. The next step was to extend the systems engineering of a single project to an entire decade of missions. That is, we wanted a true program, not a collection of missions. I had at least one example of a successful and identifiable program that I knew I wanted to emulate. The sequence of missions that started with President Kennedy's announcement in 1961 that we would land an American on the Moon and ended over a decade later with Apollo 17 had a closely linked set of technological, political, and scientific imperatives supporting the overarching goal to demonstrate to the rest of the world that the United States could in fact be the first to the Moon. This was a collection of missions that were scientifically linked, technologically linked, and possibly operationally interacting, and supported an overarching goal. This seemed like a good definition of a program. The formulation and development of these individual projects needed to be in the context of overall program objectives. Each mission would build on the scientific and technological lessons learned from prior missions. I felt strongly that such systematic thinking would stand us in good stead.

We sought to have overarching scientific goals on which to build and technologies that would be funded until they were ready to link the scientific missions through orbital reconnaissance and ground-truth, *in situ* investigations. To get to this point we actually created, in essence, a new type of discipline called program systems engineering. This is the discipline of trade studies to ensure that we itemize things at the program level, that we cross-check our requirements to ensure that the various parts work with each other, and that we monitor and balance all the risks in the reserves and margins (both monetary and schedule reserves, the extras for the inevitable unforeseen contingencies in such groundbreaking endeavors, and margins, the "guesstimates" required because you haven't done this before and are always guessing, at least a little bit) among the different missions.

In addition to the other attributes of systems engineering for the entire program, we wanted to build in robustness to mission failures—that is, any single mission would not be uniquely dependent on the next launch. Since the administrator was still Dan Goldin, who was very much an advocate for faster, better, cheaper, we wanted to continue to apply those three principles where it made sense and where it was prudent. It was also clear that our budget would not support everything the scientific community wanted to do, so the establishment of comprehensive partnerships was necessary to add additional scope and capability.

Finally, as a programmatic attribute of our new creation, we were committed to providing data to the public as soon as possible. This meant that we needed very high bandwidth to allow for images and other data returned from multiple sites, plus interpretation of complex data as quickly as possible because, after all, the public was paying for our program.

Other goals that were integrated into our overall program included the scientific goals of astrobiology, that relatively new discipline. In its current form, astrobiology had been around for only about five years at the time of our program redesign. Nevertheless, the new NASA Astrobiology Institute already boasted inclusion of perhaps a dozen lead institutions and hundreds of principal investigators and senior researchers.

A further scientific program attribute we greatly desired was rapid response to new scientific findings. This led, in part, to the pacing of the missions so that the findings from one mission could influence, perhaps, the next mission, but certainly the mission beyond that. Also, we wanted to allow both for hypothesis testing (posing a scientific question) and for serendipitous discovery (in the event we found something truly surprising).

As to the ultimate goal of human exploration, it turns out that the goals of scientific exploration and the goals of human exploration overlap to some degree, but not greatly. Each group starts from a different set of objectives that can lead to a different type of mission. We needed to develop technologies that did not exist in 2000, particularly long-life, long-range exploration tools. We needed orbital reconnaissance, intermediate-altitude reconnaissance, local and long-range mobility, subsurface exploration, and, of course, ultimately a sample return. It is highly risky to incorporate technological development while one is trying to build a space mission—

what if your technology development lags and you end up with a whole team waiting for the technology before they can continue the next step of the build? So our budgeting and investment approach was to provide new technology well before the mission start.

Another consideration in our program systems engineering process, then, was technological readiness. In putting together a space mission that would orbit or land on Mars, there are whole hosts of technologies that must be available during the three years or so that the project is actually constructed. These might be communications technologies. They might be scientific instruments. They might be launch capabilities. One of the jobs our team took on was to look in an iterative fashion at the missions and projects that were being developed and their requirements for technology and make an assessment of the readiness. There is a scale used within NASA known by the bureaucratic title of "technology readiness level" or TRL. The scale goes from 1 to 9, where 1 is the concept and 9 is something that has flown successfully in space and is proven technology. The usual breakpoint for a project manager who tends to be very conservative to incorporating new technology is TRL6. So our search through the missions and projects that were under way was to understand if we had any technologies that were being counted on that were below TRL6.

Also included in our program systems engineering was mother nature. Celestial mechanics and physics of the alignment of the planets turned out to play an unexpectedly important role in selecting our missions. The simple schematic shown in chapter 1's figure 1 provides the general case, but in fact the various factors involved, such as Earth's orbit and Mars's orbit, go through cycles that are a bit different every year. We found, for example, that the year 2003 was an extraordinarily good launch opportunity where, with a given rocket, we could get much more mass to the top of the atmosphere of Mars and down to the surface. By contrast, the year 2005 was a very challenging year in which even a large rocket would struggle to get our notional Mars spacecraft into orbit, much less to the surface. So, we folded in a whole series of constraints dictated by the physics of the planets.

And finally, as stated, we also had to pay attention to public engagement. This was not to be the determining factor, however. We were working with taxpayer dollars, and we felt that it was always important to find ways to give the data to the public, to engage the public in the mission, and to find simple and straightforward ways to characterize to the public what we were doing and at the same time be efficient and effective with the budget to accomplish the scientific and technological task at hand. If it sounds like it was all a lot to juggle, it was.

In 2000, then, with pressures from all sides—an administrator demanding unrealistic results now, a science community desperate to find out if there really might be alien life right in our own solar system, engineers eager to redeem themselves and simply design a mission that wouldn't fail, and my own experience—I knew I needed a simple, readily communicable science strategy to please the various groups. I wanted an underlying principle or set of principles that would make a coherent signal out of all the noise. I wanted a way to calmly explain to everyone from Congress to Nobel laureates why we were doing what we were doing and why it made sense. And I needed a way to talk about it all simply and compellingly.

I turned to all my advisers and set them on the path—not to come up with a specific mission, but rather to think about a strategy for making wise decisions that would last beyond the time when we were assembled in Washington responding to an emergency. I knew we were on a good path when we hit upon "follow the water." This epiphany happened in a very early review of the existing Mars Program shortly after I arrived as Mars Czar. Carl Pilcher was presenting the charts derived from the 1995 analysis "Exobiology Strategy for Exploring Mars." (Exobiology was a partial predecessor to astrobiology.) One of the viewgraphs displayed the presence of water as a crosscutting feature of exploring Mars for life. It was at this review that I noted that the notion of following the water could be a catchphrase for the program.

Thus our science strategy became known to us and to the world at large as "follow the water." The elegance and simplicity of this intuitively obvious statement stood me in excellent stead in the end, easing my explanations and justifications to all those high-pressure forces. There were, of course, underlying engineering principles that were equally important. In the complex world of space and space missions, the ruling engineering discipline is systems engineering. Every element of a spacecraft and a mission may be developed by its own specialist: structural engineers for the basic physical platforms, mechanical engineers to understand and guide the forces that come to bear, electrical engineers for power, propulsion engineers to make things go, software engineers to command it all to work—the list is seemingly endless. But the genius who oversees the entire final effort has to come from systems engineers. These folks have to see the big picture. They have to look at it all and see how it will come together and work as a seamless whole, all happening thousands or even millions of kilometers (1 million kilometers is about 600,000 miles) distant, at that. Firouz Naderi and Gentry Lee are two of the best systems engineers I have ever come across. They helped me go back to simple, driving, first principles.

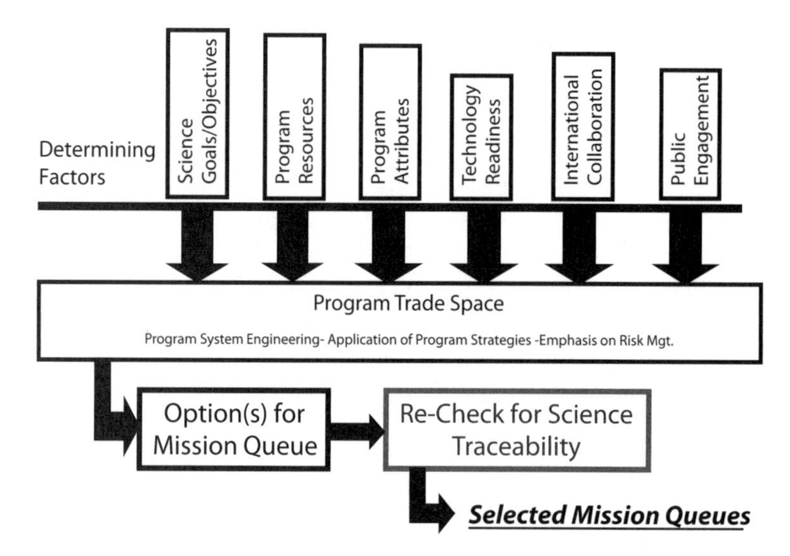

Figure 5. As described in the text, program systems engineering was developed to successfully trade and balance the many elements of a decade-long series of interrelated Mars missions.

At the topmost level on any space mission, there are three big-picture elements that all play a role. They are science, technology, and management, and in conjunction they create the program trade space—the realistic parameters within which you can make decisions. The trade space generates various options for the mission queue, which in turn is rechecked for solid science traceability. Will a given mission with a given set of instruments, and the attendant data set produced by those instruments, result in the science advances that we most want? On this basis we can then select an appropriate mission queue. I made use of the simple schematic shown in figure 5 to drive this process home.

To a lay audience such an approach may seem overly simplistic or even so obvious as to be unnecessary, but, believe me, when large, often powerful groups are tugging in different directions to have their voices heard loudest over everyone else's, it pays to have a simple process, which you hope is accompanied by an intuitively obvious graphic, that you can point to that justifies your decisions. This basic, first-principles-based, systems engineering approach was the backbone of the new Mars Exploration Program.

Next we needed an equally compelling science strategy. We had the words "follow the water," based on the fact that wherever there is liquid

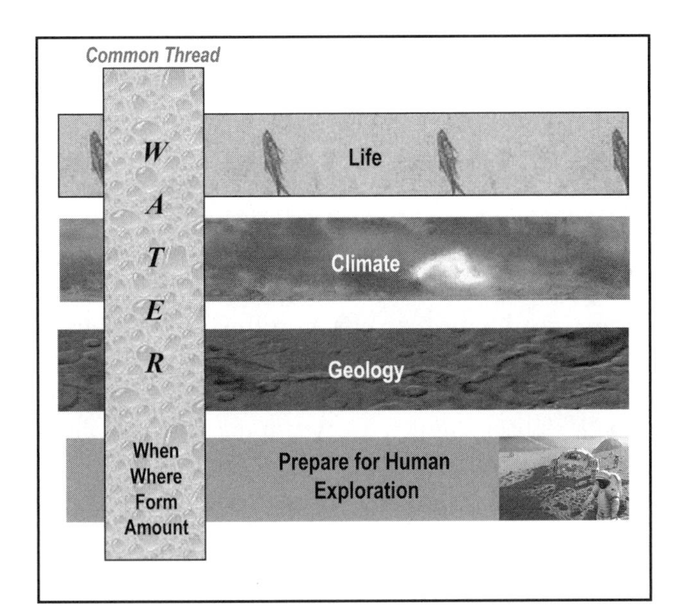

Figure 6. Growing from the Mars science community was the consensus that understanding the presence of water could serve as the common thread for our program. (Courtesy NASA)

water on Earth at any point in the annual cycle or even multiyear cycle, we find life of some sort. Surely on Mars, if there is or was ever a chance for life to emerge or survive, it would be associated somehow with water. With the help of the scientists on my team, we came up with four areas for consideration: *life*, the desire to understand if it had even been there and where, if it had survived, it might be now; *climate*, to understand the Martian cycles and how they might affect water and the water cycle on the red planet; *geology*, the stalwart workhorse of planetary studies where most of the secrets probably lay; and *preparation for human exploration*, the brass ring for many voices within NASA. The crosscutting element that affected and defined each of the four areas from our point of view was water. The questions we posed for water were: When? Where? What form? In what amounts? We were confident that by promoting and sticking to these vital questions we could vastly enhance our understanding of Mars and have a solid and justifiable Mars Exploration Program that would never again be a simple collection of unrelated missions. We summed up these notions in another simple graphic that told thousands of words.

"The devil is in the details" is, perhaps, nowhere more true than in space. A solid systems engineering strategy, underpinning an equally solid science rationale and strategy, represented just the topmost level of thinking. We had to drill down to much finer grains of thought even to get to the basic mission queue, let alone getting those missions funded and off the launch pad. One clear next step to take was to decide how to best take advantage of experience and lessons learned. We had seen how even a theoretically successful set of missions such as Viking could result in a stalled program if the results were not what everyone expected or wanted. How could you design a program that would make incremental steps that would lead inexorably to a wanted result?

From a science perspective, we approached the basic issues the way any laboratory scientist would. You start with a hypothesis. You define a likely experimental apparatus; in the case of a Mars mission, this means the spacecraft and its instruments. You take some data. You analyze the information and hopefully learn something which you integrate into your thinking, and then you loop back through the system until you get where you want to go. For Congress and the public, "follow the water" often sufficed. For the science community from whom we needed solid support, we still wanted to be simple and elegant, but we had to reach a much deeper level of detail.

Scientists are basically skeptics. They are trained for that. "Show me the data." Most scientists won't accept even the simplest of statements until you recite chapter and verse: Where did you get your information? And then they still want to see the data because they never believe that someone else analyzed it as well as they might have. For this crowd, we realized that from what we knew of Mars in the year 2000, there were literally thousands of potentially worthy landing sites on the planet. There was no clear consensus of even where to start looking for the fingerprints of life. Okay. Start with thousands of potential sites and methodically narrow that down. Get more data from your orbiters and come to a consensus on hundreds of possible sites. Analyze further and narrow the search down to tens of sites. Eventually make a decision on the likeliest spots on Mars to have supported life at some point in the history of the planet.

We pictured our approach like this: The "Seek" piece encapsulated the need to look in reasonable places for what we wanted to find, based on the best information available from orbital reconnaissance. This may seem like a given, but in planetary exploration, landing sites have historically been determined at least as much by lowered risk due to engineering aspects as by any science potential. We tend to want to land on flat, smooth surfaces near equators because that's the safest and easiest place to put a spacecraft on a planetary surface. It's a bit like the drunk in the old joke: "I dropped my keys over there, but I'm looking for them here under the lamppost because that's where the light is." We wanted to show clearly that there was a logical intellectual progression from establishing the context to wanting

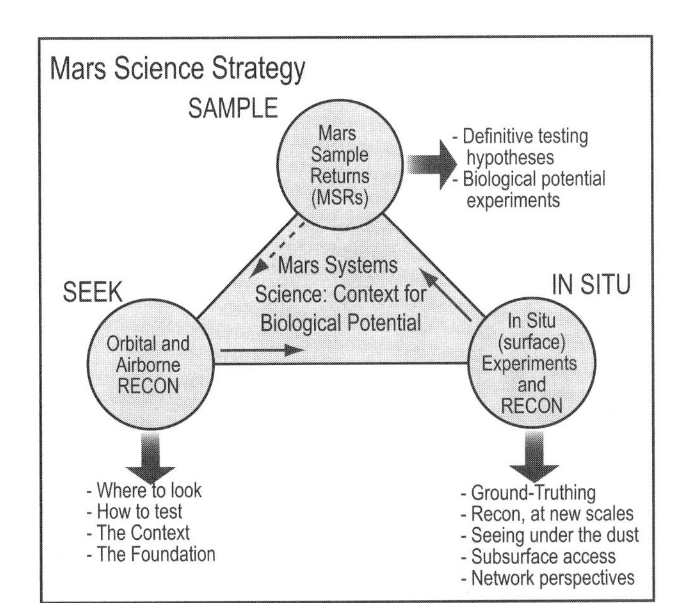

Figure 7. "Seek/*In Situ*/Sample" emerged as a logical sequence of reconnaissance, ground truth, and (eventually) sample return. (Courtesy NASA)

corroboration and ground truth. This was expressed as "*In Situ*," with options spelled out both for looking beneath the surface and for potential networked or multi-station/multi-lander concepts that had been under discussion for some years. Only after this preliminary groundwork was accomplished should one go after the third leg of the stool, we argued, which was highly focused experiments, ultimately resulting at some future point in Mars "Sample." With such clear methodology, we won over many voices who might otherwise have retained doubts about our decisions.

From an engineering point of view, we looked at the types of missions we had to deal with, coupled with our known launch opportunities. We all knew that orbiters were easier to build and fly successfully and landers were where the greatest breakthroughs in knowledge often came—"ground truth" that validated the remote sensing. We had a solid launch opportunity every 26 months. We would commit to taking advantage of every one of those launch opportunities. The one coming right up was already nipping at our heels. I was convinced that the lander was a mistake. It just wasn't ready to go, but perhaps more importantly, there wasn't enough rigor behind where it was going or with what instrumentation. The decision had been made to send the orbiter and not the lander, but for the next opportunity, things might be different.

For the 2003 launch opportunity, there was still not an enormous amount of development time, but there was some—about 38 months. We

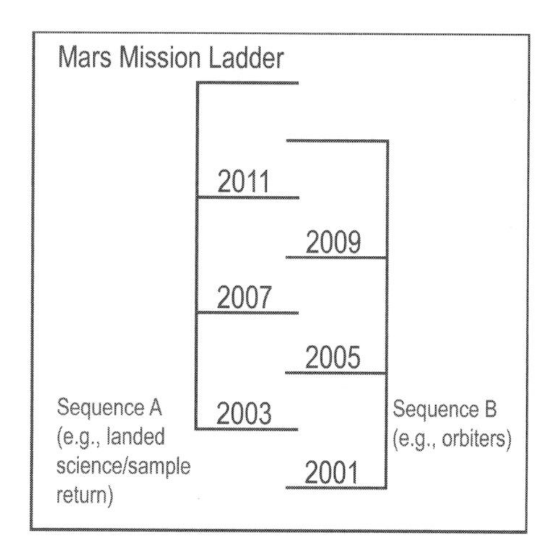

Figure 8. The "ladder to Mars" displays a sensible approach to the next decade by alternating orbital reconnaissance with landed missions.

had the successful experience of the 1997 Mars Pathfinder mission and the rover *Sojourner.* If we used every alternating opportunity and thus had four-plus years between the launches of two rover missions, we should have enough time to be able to glean good insight from the earlier mission and integrate some of those lessons into the new development. We should be getting great data from the new orbiter, the 2001 Mars *Odyssey.* That would help with site selection for a landed mission and give a host of other insights into the planet itself, including its cycles, weather patterns, and basic structures. We could take advantage of all these things by sending a landed mission during the 2003 opportunity.

As the team began thinking along these lines, I hit upon another simple, easily communicated notion of solid engineering principles that could underpin and benefit the entire program—the "ladder to Mars." The basic concept was that one set of opportunities could be orbiters, the kind of mission that provides great context and enhances the understanding of a planet as a system, and every other opportunity could be a lander that provides the corroborating ground truth and the details that advance the science in new and exciting ways.

An important element of the new rigorous systems engineering thinking in which we all became engaged might seem like the most obvious of all but may very well have represented the most important shift in Mars exploration thinking and our greatest contribution. We all began to realize that Mars itself is a system and can be, and should be, approached as that. Its geology, its windstorms, its almost-warm days and frigid nights, its melting polar icecaps, its moons, all of it represents a complex system. We could profit from thinking about Mars explicitly in this way.

Finally, we knew the importance of "buy-in" from the various stakeholders, such as the science community and the large number of aerospace contractors, and we wanted their support. We also knew that in a world of rapidly changing technologies and huge technological leaps, we wanted a way to be able to capitalize on new developments and innovations. We needed to look ten years down the road, but how good were our crystal balls? We hit upon a program element that allowed us to accomplish both of these goals at once.

There is a very popular program within NASA called the Discovery Program. This program consists of small missions that are selected through the scientific peer-review process and which may have as their objective the investigation of the Moon, the investigation of an asteroid, a tour and return from the tail of a comet, or other goals. While it does not fit my definition of a coherent, linked program with recognizable objectives, it does encourage great innovation and allows principal investigators with unique and compelling ideas a shot at getting a full mission funded within NASA. We integrated what we called the Mars "Scout" missions at a couple of opportunities into our mission queue. In this way, we provided at least a couple of open competitive proposals for the Mars community.

With the "ladder to Mars" defining the basic structure of the mission queue, alternating landers and orbiters, the next piece of the puzzle was the nature of the all-important science instruments. We were now looking at context—that is, at possible habitable environments for potential life on Mars—so what would be the best way to systematically and methodically narrow our focus to the right places, those deserving the extra cost and risk of a lander? Again, we took the simplest possible approach. We had Viking-era Mars imagery that represented roughly thousands of meters per pixel, very low resolution by Earth standards. Mars Global Surveyor, launched in 1996, had improved this resolution to 30–100 meters with some selected areas of a few meters. We would ensure that subsequent orbiters provided at least improvements to tens of meters and then 1 meter or less per pixel until we got down to centimeters and the kind of detail that would provide deep understanding. For our landers, we would similarly demand that each subsequent lander provide a real leap in ability, building each time methodically from the lessons learned from the previous generation. We would follow the water to a rock-solid mission queue.

CHAPTER FIVE

To Land or Not to Land?

From late March, when I was appointed, until early October 2000 the replanning of the entire Mars Program was ongoing, not in an easy-to-follow serial fashion, but rather in a multiplex, multifaceted, intertwined, and interconnected way. As is customary with NASA missions, there are a series of reviews between approval and launch. In the midst of deciding on the 2003, 2005, 2007, and 2009 missions I still had to attend to the launch preparations for the 2001 mission, the orbiter, Mars *Odyssey*.

As difficult as the decision to cancel the 2001 lander had been, an even more challenging and complicated decision was whether we would try to fly a mission in 2003 or not—and if so, what would it be? This decision was actually being made as we were developing the science and engineering strategies outlined in chapter 4.

Throughout my tenure as Mars Czar there were multiple planning issues proceeding in parallel, which required a great deal of juggling and simultaneous decision making. The appendix lists the multitudinous issues and actions I needed to address. The issues had varying degrees of urgency, but I knew that everything needed to be concluded in time to match the upcoming budget cycle. In addition, there was the unforgiving celestial mechanics cycle of the 26-month launch window from Earth to Mars. That window of only 20 days every two years or so was the single greatest pressure that everyone on the team felt and is a truly unique constraint of planetary science missions.

The original decision to launch a Mars orbiter as well as a lander in 2001 was made very quickly, but the process of planning what to do in the

2001 opportunity had, in fact, begun even before I was named as Mars program director.

I arrived in Washington, D.C., and took up an apartment in early April 2000. I had met with Dan Goldin on March 11, given him my "Yes, I will do this" answer within a couple of days, and taken off for Washington, D.C., soon after. To make all this happen on such a short schedule was an exercise in urgent planning and rapid deployment. My wife and I took a very quick trip to D.C. and looked at a number of possible apartments, keeping in mind that the government's compensation for a relocation like this was minimal. On our taxi ride out to the airport on the afternoon of the final day, we asked the driver where there might be places to live. To our pleasant surprise he had a good answer. He said the greatest place is Alexandria, just across the river from D.C., and if you get the right apartment, it's near a Metro line. If you've spent any time in Washington, you know that the Metro, the sprawling subway lines that serve the nation's capital and its suburbs, is the very best way to get around town. The surface streets are confusing and clogged with traffic during most of the regular hours of the day—the subway makes that moot.

Of course, this rapid move from Burlingame, California, to Washington, D.C., also included packing up a very aged cat along with my wife and our household goods, moving everything out of our house to storage, and quickly leasing the house to what turned out to be a great young business couple who had moved to town as a result of a corporate consolidation. I moved first, to establish a beachhead in our Alexandria apartment. Susan took care of the logistics—no easy set of tasks—of packing, storage, and house leasing and arrived in Washington about three weeks after I did.

As I assembled my staff and looked at the immediate decisions and at the unchangeable launch window clock, it was very clear that the decision to fly or not in 2003 had to be made nearly instantly. Ordinarily a new mission will receive one to two years of intensive study followed by three to four years of development. The launch window in 2003 was only 38 months away, which meant a very rapid development time, no matter what was decided.

My notes show that discussions about the mission decision for the 2003 launch opportunity began at least as early as April 11, shortly after I arrived. We had established a program replanning executive council for the Mars missions that included not only members of the science organization, but also people from other NASA groups who had a strong interest in the program, including the Human Exploration and Development of Space (HEDS) enterprise responsible for overseeing all human exploration. All

NASA organizations go through periodic name changes as administrations change. As a result, that group no longer exists, but clearly, the functions are still there.

The 2001 Mars lander that I canceled had planned to include experiments that were directly related to the question of future human habitation. One of the experiments was to directly test whether or not the use of the Mars atmosphere to create oxygen and/or energy was a feasible approach. It's actually quite common within the space world, due in part to the significant expense of developing anything for space, to repurpose concepts, plans, and even hardware that have been partially or fully developed. Not surprisingly, the representatives of HEDS wrote me within a day or so of our discussions on the plans for 2003 to endorse the idea of repositioning their experiments for that mission.

The summary of the discussions that we held, recorded in a draft document on April 11, laid out some top-level issues. We wanted to use our newly developed idea of following the water. We had to assess the schedule, budget, technical readiness, and risk for the 2003 launch, and we had to integrate all this into a long-term Mars Exploration Program. Such an approach was one of the themes that I think became a hallmark of the restructuring of the program. Each mission decision was made with an eye toward how it contributed to the long-term scientific goals of understanding Mars as a system. We wanted to avoid any sense that any one mission would be a one-shot opportunity disconnected from everything else. We decided in our meeting on April 12 that the Mars 2003 decision had to be made by the first week in May. We determined that a small group would visit JPL in Pasadena before decision time and that the outcome of this review would be turned into a recommendation to the associate administrator, Ed Weiler, and ultimately to the administrator, Dan Goldin. We were, it seemed, developing strategies for our strategies, but we needed to be methodical, and we knew it. It paid off in the end.

The science organization at NASA is driven by the peer-review process: Proposals are solicited from the entire science community, put through a rigorous review, and then the instrument or mission that best satisfies the top-level requirements is selected by, usually, the associate administrator. It's a fairly time-consuming process and adds at least a year up front to any given endeavor. In this way, however, the community is involved, invested, and confident that plenty of folks have had a chance to weigh in on important decisions, and, often, innovations emerge. In our discussion on April 12, it was quite obvious, though, that this process, which would involve a

Figure 9. An ebullient Dan Goldin learning that *Odyssey* was safely in orbit around Mars. Left to right: Goldin, his grandson, Tom Gavin, Matt Landano, and Hubbard. Far in the background the president of Caltech, David Baltimore, can be seen.

NASA "Announcement of Opportunity" and over a third of our available time, was infeasible with only 38 months to the launch of the 2003 mission.

Instead, we decided that we should focus on instruments, investigations, and flight hardware that had already been selected through some peer-review process or could be rebuilt from previously flown missions. As I say, this is actually not that unusual in the space business where everything is hard won, so no one wants any effort to go to waste. At that point on April 11, we were thinking that if the 2003 mission were a lander, it would probably use instruments built and tested for the 2001 lander or perhaps utilize rebuilt MPL instruments in some combination. If the decision was to fly an orbiter in 2003, our thinking was that instruments that were aboard the MCO, already selected through peer review, could be rebuilt as well. We had also begun at this point to think about the possibility that the science community could propose an entire small mission following the lead of the Discovery model. The concept of a "Scout" program was already emerging at this early date.

In addition to assuming that we would find instruments or investigations that had already passed through peer review or could be rebuilt from previous missions, we made the decision that, given the time urgency, we had to work primarily with JPL, which has the responsibility for implementing most of the planetary missions that are flown by the science organization. There simply wasn't time to involve others even though they might be unhappy about being left out. We began to review and scrub the guidelines for selection of the 2003 mission on an almost daily basis. As the guidelines evolved between the first discussions on April 11 through the meeting itself on May 4 and 5 at JPL, we continued to emphasize that the instruments and the science had to be ready to go. In a practical sense, this meant that the instruments should have already been built and environmentally tested—frozen, heated, shaken, spun, and subjected to all the other grueling tests we perform on space hardware to ensure it will survive the punishing trip to another world. What we wanted was essentially "off-the-shelf" hardware.

Although we had decided that we must work with JPL and the existing planetary science infrastructure to meet such a demanding schedule, we also wanted to reach out to as much of the community as we could so that we were as inclusive as possible on a short timescale. It is interesting now to reflect on the draft agenda for the 2003 baseline mission review we prepared and sent to Earle Huckins back at NASA HQ on April 18. The agenda at this point contains no reference to Steve Squyres's Athena payload packaged in a Pathfinder system. The draft contains, after some introductions, a discussion of the lander element, primarily left over from the 2001 lander mission; an orbiter, attempting to recapture much of the MCO data; a small mission concept known as "Scout"; and an element for telecommunications infrastructure.

In the interests of being careful and complete, I reached out to my old friend Tony Spear, retired JPL Pathfinder mission project manager. On April 27 he responded to me in an e-mail saying that, based on Pathfinder experience, he concurred that there was no time for a major new development. He pointed out that they had had 19 months in a pre-project phase to do the planning and early airbag tests, and that it had taken almost 42 months before they had the airbags ready for launch in the spring of 1996. His note went on to suggest a series of options for 2003, such as modifying the 2001 lander that was in storage, creating an orbiter with Mars Observer-class instruments, and a re-flight of Mars Pathfinder with a *Sojourner-2*, but he also asked what would the mission value of such a mission be? Then he concluded his note to me by saying that we would have to make this decision quickly because of the launch schedule pressure, without having had a chance for a thorough peer review by a wide audience. He was reinforcing my worst fears, in fact. The note suggested that one strategy might be to have no lander at all in 2003 and, instead, focus on landing-site selection with high-resolution atmospheric and surface measurements. Finally, he noted that one of the most successful undertakings of the 1990s was the Discovery Program of small, principal-investigator-class, peer-reviewed, and peer-selected missions. The Lunar Prospector mission that I'd worked on several years before was the first such fully peer-reviewed example, and it was quite successful. The last sentence in Tony's note says that maybe there should be a Mars discovery program. Great minds think alike, apparently.

As we got closer and closer to the May review dates, the executive team at NASA HQ, working with the JPL Program Office, produced the charge to the board. We had arranged the JPL May meeting to have a board of review consisting of all of the stakeholders in the Mars Program, chaired by me with Earle Huckins as deputy. We would look at all the options that could be studied in the short span of a month. The final official title of our note to everyone was "Mission Options Assessment Review." In Washington and within NASA there is a hierarchy of responsibility and accountability. The responsibility for selection lies with the associate administrator or, if it's important enough, the administrator. So calling our meeting an "assessment review" was a way of making sure that we didn't give the appearance that we had already made a selection in advance of the most senior people looking at the recommendations. In the minefield of federal politics, stepping on your boss's toes or, worse, your boss's boss's toes is particularly frowned upon.

The content of this Mission Options Assessment Review consisted of the statement of background and purpose, which reiterated that in the wake of the loss of MCO and MPL we were restructuring the entire program and, given the schedule urgency for a 2003 opportunity, it was important to conduct a fast-track study. We were dotting every "i" and crossing every "t" we could find.

The assessment guidelines laid out the rough priority order of what we were looking for in the next mission. As I stated in the write-up that I'd distributed to Earle Huckins and Ed Weiler and eventually everyone else, the selection criteria were in roughly priority order, with risk evaluation and science merit having more weight than strategic integration, feed forward (laying the groundwork for future missions and efforts), and public engagement.

Within risk evaluation, the top item is technical risk. We had to consider the technical maturity of the options and feasibility of successful development, not only in general, but especially considering the short schedule. What were the technology requirements and heritage of the hardware and software? We had already seen how software can really trip up a mission. What were the "long lead items," those things that took the longest and therefore represented the greatest risk to the schedule, even when relatively well understood? What was the risk of the technical design, the science instrument development, or other schedule risks? What could be done, and done well, with a two-month study phase and 36 months or less for development?

After risk evaluation, and of very close weight, came the science merit. We had to ask ourselves how the proposed science payload supported the objectives of the Mars Exploration Program. Will we recover the high-priority science from the two lost missions? Is the payload consistent with the highest-level goals of seeking life and climate resources, with life being the first among equals? Do the objectives of the proposed option "follow the water"?

We then had to make some assessment of how the proposed option for 2003 supported and integrated with a 10- to 20-year Mars Exploration Program. Would the proposed option appear to be a dead end or would it provide a logical building block for extension?

Finally, in a not insignificant way, we had to evaluate the public excitement for the proposed option. After all, we were spending taxpayers' money, and we wanted to be sure that we engaged them in our effort. For example, we said, "Is the science implementation likely to be engaging? Does the proposed option appear to be repetitive or are there elements of novelty and probable surprise?"

These final criteria I sent out May 1 to the groups that were preparing for presentations on May 4–5. Interestingly enough, on that day, I received an e-mail from Steve Squyres in which he said that he urged all of us, including JPL, to take a careful look at a mission concept that would deliver a capable rover to Mars using the Pathfinder airbag system. He went on to say that, in his view, if there was to be a mission in 2003 to the surface of Mars, it would likely come down to a choice between the rover airbag concept or modification of the 2001 lander. He concluded by saying that he believed the instrument suite that his team built for a mobile platform would be able to perform the science that it was originally proposed for. The key to the water and life question is in the rocks, he said, and getting an instrument to a rock requires mobility. Throughout my career, I have seen scientists be tireless advocates for their concepts and their instruments. When they are, even though there are examples of the concept requiring decades to be successful, they often prevail. Squyres was and is no exception.

And so it was that at the last minute, as we developed the agenda for the meeting that first week in May, there was an additional one-hour presentation added. After lunch and after hearing about the restructured 2001 lander, a small Scout mission, a telecommunications element, and an orbiter, we heard from Mark Adler of JPL about a lander in an airbag, building on the success of Pathfinder. We also heard an assessment of the relative merits, risks, launch conditions, and other technical factors that would influence the 2003 mission. All told, there were perhaps twenty or so different options, combinations, and permutations presented for us to consider.

During the meeting there were, of course, a lot of opinions expressed. In my notes from this meeting, after listening to all the debate and discussion, I drew conclusions for each concept. For the Scout, the schedule simply seemed to be too tight. There were no simple surface instruments of a 5-kilogram (10-pound) class developed, and the science didn't seem to meet the minimum threshold needed to justify developing a full mission. Much more thought and the engagement of a much larger community would be required to get great ideas. Developing a dedicated telecommunications orbiter, while necessary from a program perspective, didn't seem to be coupled to the overall requirements for the 2003 mission. The orbiter, while containing some recovered science from the lost MCO mission, raised questions on the feed-forward nature of the project. The presentation team clearly had the Mars sample return in mind, and the definition of the actual project was not yet crisp. The revisited 2001 lander still had all of the questions about landing success that were raised in the previous review that had led to the termination of that mission. The concept of putting the science payload on the mobile platform of the Mars Pathfinder system had a science appeal, but many questions about cost.

As I summarized everything that I'd heard, it seemed to me that the community had been in a reactive mode since the loss of Mars Observer in 1993. We needed two successes in a row. We needed adequate resources to invest in technology and to plan and retain flexibility for 2005 and beyond.

I felt that my role, while clearly that of program leadership, was also to listen carefully and assess and integrate the comments of those people whom I had respected and worked with for so many years. My final notes at the end of the two days were that there seemed to be a consensus that a small Scout-class mission would be useful for, perhaps, 2005 and beyond, not 2003. A number of the options we had been presented still retained the previous era's directive to fly orbiters and landers at every opportunity. After the discussions of the risk and the cost, focusing on a single element or a single mission type for 2003 emerged as the best future path. At least we had accomplished some reliable decision making.

One of the messages that came through very clearly in this meeting and on the subsequent day was that a critical issue was to determine if we needed mobility to carry out science. Can you do great science from a single landing spot with relatively sophisticated instruments, or do you really benefit from the trade-offs of smaller, simpler instruments that can be carried on a rover with the capability of moving those instruments from rock to rock and place to place? I distinctly remember two very senior and respected members of the community, Ron Greeley from Arizona State University and Mike Carr from the U.S. Geological Survey, making eloquent statements about the need for mobility. Ten years later, with the incredible successes of the two Mars rovers *Spirit* and *Opportunity*, it seems hard to imagine that there was a time in the Mars Program when the need for mobility was hotly debated. Carr pointed out the incredible success of the Viking mission but noted also its shortcoming—that only those objects within the reach of the arm attached to the platform could be tested.

The other consensus I noted was that support for multiple elements—some sort of multisatellite configuration with multiple spacecraft—was very thin. This assessment was consistent with the consensus that a single-element focus was the best choice for 2003. I observed that the sense of the group was that the scope of a telecommunications satellite had to be revisited if it was to be implemented at all. It was not clear how you would do a communications satellite, fit it with both the scientific and programmatic goals of the future, and stay within the proposed budget. Finally, I got the sense that we, as a community, had not really understood the distinctions between the Pathfinder airbag landing technique and the Viking-style retro rockets being proposed for a new version of the MPL for 2003.

The results of this two-day meeting and the month of study that had gone into it, then, were distilled into three distinct possibilities: a static lander derived from the 2001 mission and MPL, a science orbiter of Mars Global Surveyor class, and the dark horse that emerged at almost the last moment, a science payload on a mobile lander, packaged inside a Mars Pathfinder envelope. I should note that although the charge to JPL and the science teams gave them only a month to produce these initial presentations, the mission concepts, the instruments, and the science objectives were ones that had been studied and evaluated by the community over a period of many years. I felt fairly confident that the options we were consid-

Figure 10. During the development of the 2003 mission at JPL, a model of the 25-pound Pathfinder rover (left) engaged in a chase around the clean-room floor with the 384-pound MER rover. (Courtesy NASA/JPL-Caltech)

ering had been given deep thought and had community consensus behind them. When I returned to NASA HQ, I quickly sketched out a comparison of the three choices and faxed it to Firouz Naderi at JPL. The handwritten fax sheet in figure 11 shows our thinking.

Firouz and I were already beginning to sketch out the possibilities where "follow the water" might lead us for specific mission concepts in 2003, 2005, and 2007. In addition, the risks and rewards of the three approaches for 2003 were becoming very clear, so in parallel Firouz and I, with the NASA HQ team and the JPL team, set two efforts in motion. The first was to create three study teams—the Pathfinder-style, mobile lander team; the Mars Global Surveyor orbiter team; and the MPL-like team—and give them more time to evaluate the risks, costs, and schedules for each 2003 option. Our second effort was to further flesh out the 2005 opportunity, the 2007 opportunity, and beyond. We knew that to do this effectively we needed to broaden our reach and broaden our scope to an even larger segment of the Mars community.

The final chapter in the selection of a rover for 2003 came on July 13

MARS 2003
OPTION COMPARISON

OPTIONAL FORM 99 (7-90)
FAX TRANSMITTAL | # of pages ►
To Firouz Naderi | From Scott
Dept./Agency | Phone #
Fax # | Fax #
NSN 7540-01-317-7368 5099-101 GENERAL SERVICES ADMINISTRATION

MOBILE LANDER PATHFINDER	SURVEYOR SCIENCE ORBITER	STATIC '01-DERIVATIVE LANDER with ARM
• ~1000 m TRAVERSE to investigate ROCK MINERALOGY/chemistry	• Multi-INSTRUMENT "MGS" class ORBITER to follow the modern (atmos.) and ANCIENT water	• COMPREHENSIVE "fixed" in situ investigation of surface materials (geochemistry, mineralogy etc.)
• SEARCHES in situ for Aqueous minerals (water-related)	• RECAPTURES MCO SCIENCE, Extends MGS and MARS 2001, AND FLIES first high-resolution NIR imaging spectrometer	• Largely recaptures 2001 LANDER SCIENCE (no Marie Curie rover or ISRU experiment)
• Uses repackaged APEX instruments from '01 LANDER	• SEARCHES for aqueous minerals from ORBITS and assesses LANDING SITE hazards	• FIRST DESCENT IMAGING with 10 cm to 10 m resolution
• PATHFINDER ENTRY, DESCENT, LANDING SYSTEM	• MGS, M'01 ORBITER heritage	• First analysis of surface dust via microscopy & wet chemistry
• NO direct HEDS experiments	• MCO, MGS instrument heritage	• FIRST surface RADIATION ENVIRONMENT ASSESSMENT
• DEVELOPMENT RISK of NEW MEDIUM-SCALE ROVER SYSTEM	• DEVELOPMENT RISK of ≤ 100 m resolution NIR Imaging Spectrometer	• RISK of LANDING 3-legged Vehicle w/ 50 cm clearance on MARS
• POSSIBLE POWER ISSUES given SOLAR SOURCE	• SCHEDULE RISKS ASSOCIATED with instrument acquisition	• LANDING Safety could (WILL) drive landing site selection
• RAPID RE-ESTABLISHMENT of TEAM REQUIRED		• NO MOBILITY - may not be able to reach a ROCK
		• ROBOTIC ARM for subsurface access

Figure 11. A handwritten summary of the three mission options Hubbard faxed to Firouz Naderi in spring 2000 as they were rapidly deciding what project to recommend for the 2003 launch opportunity.

and 14, 2000, the date of the decision meeting. By the time we met in July, the three teams had accomplished almost two months of further study. The launch vehicle and celestial mechanics teams by this time had calculated the launch windows with the greatest degree of precision possible. The best that we could hope for was a July 2003 launch opportunity from Cape Canaveral Air Force Station (CCAFS), the launch area in Florida that has the responsibility, along with contractors such as Boeing, for the launch of missions that do not involve human spaceflight. These calculations confirmed that we were now a mere 36 months away from sending another mission to Mars. I cannot emphasize too much how stressful this demanding schedule is. It is the single most unforgiving part of planetary science. Missions that orbit Earth, missions that trail Earth and look out into the universe, even missions that go to the Moon have many launch opportunities. But to go to Mars or Venus or Jupiter or any other body orbiting the Sun, the launch window is limited by the alignment of the celestial bodies and physics itself.

We had selected a review board very similar in composition to the group that had helped us evaluate the twenty or so concepts two months before. With some additions, like Gentry Lee and Jim Martin (the retired Viking project manager), the group we gathered at NASA HQ on July 13 and 14 was very much the same.

As with the review in May, we set out a statement of purpose and guidelines. Once again the charge to this board was very similar. The distinction was primarily in the very first paragraph, where it said that this was the second and final step in the review and assessment of what had narrowed to two proposed missions for the 2003 planetary launch opportunity. Upon completion of the review, the Mars program director would present a recommendation to the associate administrator of the Office of Space Science (Ed Weiler). As before, we said that risk evaluation, science merit, and, now, cost had more weight than strategic integration, the feed forward to future missions, and public engagement.

By July our NASA HQ and JPL teams had taken a very close look at the two landing techniques, retro rockets and airbags. We went back through the rationale for deleting the 2001 lander and looked at the proposed derivative of this mission for 2003. As we assessed the potential risks that were identified with using a three-legged retro-rocket lander versus the apparently robust Pathfinder airbag technique, it became clear that our group did not believe enough risk had been retired to pin the hopes of the 2003 mission on the MPL rocket-style idea. As a consequence, by the time of the July 13 and 14 decision meeting, we had narrowed the possibilities to either a science orbiter in the Mars Global Surveyor class or a mobile lander in the Pathfinder airbag envelope.

By this time the mobile lander in an airbag had acquired the name Mars Geological Rover and the orbiter was characterized as Mars Surveyor Orbiter, so the acronyms were MGR and MSO. The studies were conducted by both JPL and Lockheed Martin in Denver, the company focusing on the orbiter. We held a detailed review for both concepts, one on July 6 and 7 and the other on July 10 and 11. I should note that while all of the presenters were top-notch professionals, I was particularly impressed by the MGR leader—Pete Theisinger. At our reviews, in addition to the criteria of risk, science, costs, and so forth, we had independent cost estimates presented by Science Applications International Corporation (SAIC), a company with a long and respected history of consulting in the space business. There was, of course, always the possibility of doing nothing in 2003, and this was discussed at the decision meeting on July 13 and 14 as well.

The science assessment was that both missions were high value, although they addressed very different goals and should be flown as soon as possible. Both would "follow the water," MSO from orbit and MGR on the surface. From a top-level science viewpoint, there was no discriminator. The cost analysis showed that both concepts were of approximately the same value. This was, of course, assuming that the Pathfinder-type rover would be a near exact repeat from the previous mission and that the Mars Global Surveyor spacecraft could be repeated as well. The technical assessment was that both missions could be developed for 2003, although any surface mission, as we have seen, inherently carries more risk. The landed-mission risk was deemed to be acceptable in light of the breakthrough science potential and the logical progression of surface science capability. Mike Carr, in particular, noted that if we did not fly a lander in 2003 using what we had learned from Pathfinder it would be more than ten years between landed missions.

As we discussed the options, it became clear that there was very little sentiment for doing nothing in 2003. It was in the interest of the NASA science community, and, indeed, the United States, to return to Mars not only at the 2001 opportunity, but at the 2003 and subsequent opportunities. We went around the room and asked each individual to vote. I remember Jim Martin, in particular, making the argument that were he twenty or thirty years younger, he would volunteer to be the project manager and guarantee that he would bring the lander in on schedule and on budget. There is no lack of passion among space professionals. This was one of the few times that I observed Jim and one of his closest colleagues and disciple, Gentry Lee, disagree. Gentry actually voted for the orbiter. Jim made compelling scientific, technical, public-engagement, and other arguments for the lander.

In the end after the vote, I found myself in the position of having a nearly equal set of recommendations for both missions. When I reported to the group that we had a virtual stalemate, one of my HQ colleagues commented, "Well, then you can do what you want." I took both a figurative and literal walk around the block to think about that.

The discriminator in the decision was derived, ultimately, from program context. I recommended that we fly MGR in 2003 and fly something like the MSO orbiter in 2005. My reasoning was that the really optimal next-generation orbiter instruments were not ready for 2003. We needed much higher resolution and imaging. Second, we needed to understand the Mars surface and the science of the surface before committing to a sample return. Our surface understanding was far less mature than our orbital sensing. We needed ground truth for validating orbital measurements. Next, flying a lander followed a logical path of science and technology development. We would be able to test extended surface operations with a true science package and the new rover technology. Pathfinder had been, technically, a technology demonstrator and so was intended explicitly to lay the groundwork for just such a follow-on mission as we were contemplating. I believed in the robustness of the airbag concept. I had been its original advocate. Finally, for 2003 the orbital mechanics were very favorable for a lander and this would not be true again for almost fifteen years. The orbital mechanics were not favorable for a solar-powered science lander in 2005, and this decision would give us more time to develop the alternative, a radioisotope power supply for 2007 or beyond.

As had happened in early May, during the deliberations on July 13 and 14 Steve Squyres was present as an advocate for his Athena science payload that was proposed for the lander. At a break Steve approached me in the hallway, and as he had earlier, made the point that he personally believed that the scientific payoff of having any mobile science package on Mars would be extraordinary and that the public engagement would be terrific. Steve is a persuasive guy.

Within a day, Jim Garvin and I visited Ed Weiler with the recommendation that we fly a lander in 2003. Ed was skeptical, as usual, and it was only by appealing to the program context, the value of landed science, and the just emerging idea of providing, first, orbital context and then ground truth that we were able to sway Ed to our point of view. Finally, though, Ed not only accepted but embraced the idea of a rover in 2003. We had done it. We had a consensus and a decision. We all felt great and more than a little relieved. The project manager in me saw light at the end of the tunnel and was confident that a rover with a well-developed payload and a proven, robust technology for entry, descent, and landing would fit within the existing budget and have a terrific shot at being successful. Of course, I hadn't bargained on Dan Goldin.

CHAPTER SIX

The Rover Becomes Rovers

Many people probably believe that for space missions, building the spacecraft takes all the time. However, in my experience, we often understand well how to build spacecraft, and it is the scientific instruments that are the slowest to develop or the last to arrive. This is not because the scientists don't take the schedule seriously. It is because the scientists and engineers supporting them are almost always pushing the state of the art. These teams are trying to measure things around distant worlds that have never been measured before. Still, for our newly decided-upon Mars rover, we had a real strength in an already developed science payload, the Athena payload, complete with its tireless and effective advocate, Steve Squyres. Of course, having developed and pushed for the Pathfinder airbag concept, I truly believed in it. Bouncing onto an unknown, uncharacterized alien surface may seem strange or even comical, but it's an incredibly forgiving way to land when you have no idea where the next boulder might be.

On July 16, 2000, Jim Garvin and I met with Ed Weiler to give him the results of the meeting we'd held just days before and the decision I'd come to, which Jim shared, that we needed to go to the surface in 2003 with a Pathfinder-like mission that we were calling the Mars Geological Rover (MGR). Ed was initially very difficult to convince. He is well known for adopting a very skeptical attitude toward anything that has significant risk attached to it. This is probably because of his experience with the Hubble Space Telescope. We all remember that when the $2 billion telescope launched into orbit, it turned out to be out of focus and ended up requiring a costly retrofit. As the Hubble program scientist, Ed had had to explain and work through the problems that Hubble had before it became an operational telescope (now doing its job, as we know, superbly). Space is a tough business.

I think the reason that Jim and I were eventually able to prevail was because of a combination of the science readiness, the celestial mechanics, and the fact that I was beginning to develop a sense of the program architecture. In the emerging program sequence where we were trying to accomplish a decade-long science investigation, it was starting to become clear to me that to put resilience in the program, we really needed the alternating approach of landers and orbiters. Having a lander in 2003 after launching an orbiter in 2001 meant that we could start to adopt an overall approach of orbital reconnaissance followed by surface ground truth. Over the next several months, until we presented the entire program in October, this "ladder to Mars" concept became a very important part of the overall program systems engineering concept and, once fleshed out with actual missions, represented the whole new Mars Exploration Program.

After a great deal of discussion in which we presented the pros and cons, the celestial mechanics arguments, the budget arguments, the science arguments, and both the advantages and the concerns about a Pathfinder package with the Athena payload, Weiler agreed with us that it was time to go back to the surface with a lander.

Shortly after this meeting, Jim, Ed, and I went to see Dan Goldin to give him the rationale for the lander in 2003. As the administrator, Dan had the last word. A typical meeting with Dan Goldin is that you get to brief your first chart out of your package of ten or twenty, and then Dan starts asking questions and going off on his own monologue. You may or may not ever get a chance to actually make your case. This meeting was no exception. We had only just presented the conclusion that a rover in 2003 was the right thing to do when Dan challenged us with the question, "What about two rovers?"

At times like these, you really have to think on your feet. After gulping some extra air, I could hear my unbidden thoughts: "The man is crazy! . . . This is the kind of off-the-wall thinking that got us into this mess in the first place! . . . Well, at least two identical vehicles isn't quite as crazy as two very different spacecraft would be. . . . He's the boss, and I can't just blow him off. . . . Help!" A brief discussion ensued in which I pulled some cost numbers out of the air, assuming we could reproduce the lander rover package for maybe 50 percent of the initial cost, and then we had to add another launch vehicle and a cost delta or two.

I should explain that in the space world, there are a few terms that are so ubiquitous and useful that everyone quickly adopts them. *Delta*, the Greek letter (Δ), in mathematics basically stands for "difference" or "change." Its most common use in space parlance is "delta-*v*," or change in velocity, as in the delta-*v* that is required to get out of Earth's gravity well and into orbit, roughly 28,000 kilometers per hour (17,000 miles per hour). Engineers use it all the time to speak of a host of differences that crop up due to the fact that if you change one parameter of, say, a flight instrument, such as its mass, there will likely be a waterfall of other things that change as a result. For a manager, the deltas of concern are usually schedule and budget, which inevitably interact and affect one another. Dan told us to go off and evaluate the two-rover possibility and come back to him as quickly as we could with a review of the mission elements to see if there was any piece that was near failure mode—that is, extremely risky—and even though we had confidence in the Pathfinder package, to treat it as if it was a new design. On the one hand, we were dealing with a potential new mandate—"give me two instead of the one you've been working toward"—that was enormously risky by definition. On the other hand, we were being told to be supercautious in our approach. The boss is still the boss, so we went off to examine the new situation.

The next meeting occurred on July 25. We went back to Dan Goldin with a list comparison of the Mars Pathfinder mission and the 2003 Mars Geological Rover mission that pointed out all the things that were the same. The Delta II rocket, the designated launch vehicle, is highly reliable. The risk of getting to Mars was equal to or even less than that of Pathfinder because Pathfinder had shown that the entry, descent, and landing (EDL) system, including the airbags, worked. The risk of getting the rover off the platform and moving across the Martian terrain was less than Pathfinder because not only had we already done it, but the new rover would be more capable.

The risk of achieving the science lifetime of a mission is not in just getting where you want to go, it's in the electronics. We felt this could best be addressed by plenty of inspection and test, especially for the power and thermal control systems. Mars gets very cold at night. The risk for the package of the vitally important surface instruments—not the cameras, but the ones that would actually interact with the Martian rocks—is in the deployment of a mechanical arm that allows an instrument to reach an identified rock. Again, we thought that we could address this by proper inspection, test, analysis, and the use of a mechanical testbed combined with a thorough, independent review. I had successfully used similar strategies for Lunar Prospector and had seen the value of a rigorous test program.

So we summarized the risk by saying that it was less than or equal to that of Pathfinder, and we could lower the risk associated with the mechanical arm deployment.

We also addressed the risk comparison of developing and operating two landers, rather than the single one originally envisioned. For science operations, the teams of scientists are backed up by software and various other sorts of engineers who provide the brains and heart of a mission, so two separate rovers would require two full science operations teams. After some discussion, we concluded that this could be viewed as a minor risk or even possibly a strength. In the assembly, test, and operations, the old-timers who had developed the two Voyager spacecraft (launched in 1977) and the Pioneer 10 and 11 spacecraft (launched in 1972 and 1973, respectively) saw this as a great strength. They pointed out that you would have a hardware-rich program with lots of flexibility that provided its own reduction of risk. In essence, you'd have plenty of flight spares, and extra options. These "graybeards" turned out to be absolutely correct. As the two Mars Exploration Rovers (MERs) were assembled at JPL, if the first rover, MER A, ran into problems you could go and take pieces off of MER B, and vice versa. That way, you kept the entire program going forward and avoided periods of potential inactivity. In design and development, as long as we didn't change anything, you would have only one design, so that was not a risk issue. There would be a minor risk in contracting out the cruise stage, a necessary but well-understood part of sending a spacecraft across the solar system, but not a large one, we thought. As to the launch vehicle, it was the same Delta II, which was not an issue. So, in summary, we surprised ourselves by coming to the conclusion that the risks were minor or, in some cases, having two Rovers would actually give us a better chance to succeed.

Looking at the overall science return, the consensus of our team was that with just one rover in 2003, you still had brand-new science versus no previous science, so this was a one rather than zero calculation. It would be new territory—the first mobile geology lab. This would be a real opportunity to "follow the water" on the surface, where it really counted, using mineralogy at new scales. A single rover would provide that. With two rovers, though, we could follow two science threads. We would have double the chances of getting to a really interesting location. Two rovers in 2003 meant potentially twice the science with added resiliency because you can never tell what Mars will throw at you. So, our bottom-line position to Administrator Goldin was that we were confident we could successfully land and operate a single rover as proposed. The science potential for that single mission was very high. However, two landers increased the science return and offered extra resiliency for the total mission. Finally, there was the mathematics of probability. If there was, say, an 80 percent chance of a single rover succeeding, and if the failures were random (not a systematic flaw), the probability of getting at least one rover success went up to greater than 95 percent.

Goldin said, "Let's do it. Let's have two rovers." Goldin chartered Ed Weiler and me to go off and figure out the details, get JPL's commitment, and give him a firm cost estimate that he could use to raise the funds.

There was a regular meeting of something called the Capital Investment Council (CIC) between this July 25 discussion and when we ultimately got the final okay for a press conference on August 10. The CIC was chaired by the deputy administrator, who was then retired, Marine three-star general Jack Dailey. Goldin routinely delegated the chairmanship of the CIC meeting to Dailey. However, at this particular time, Goldin did show up, and he made a strong plea for two Mars rovers. He made the point that two Mars rovers would help human exploration as well as scientific exploration, and he went around to each representative of the HQ organizations and demanded an answer. To say this was unusual is not an exaggeration. I met with Goldin outside the room of the CIC just before this discussion. Dan got right up in my face, pointed his long finger at my nose, and said, "Are you absolutely sure that we can do this for the amount of money you quoted?" Which, as I recall, was about $700 million. By this time I had commissioned two cost estimates but certainly had not had the time to thoroughly understand every possible contingency, so I added 20 percent to the highest estimate. Nevertheless, I said "Yes" or "I'm absolutely sure." Sometimes you just have to play to win. As arbitrary as all this might sound, with exploration, an inherently audacious business, one in fact is always guessing a bit. Luckily, it turned out I was right.

From this point until the press release on August 10, there ensued a frenzied period of analysis, discussion, letter writing, and hand wringing. Firouz Naderi at JPL was called to look at the new concept. I asked Gentry Lee, the man who had voted for the orbiter, to review it carefully. It's one thing to assert that you can do something, but once you have made the commitment, you had better do everything in your power to follow through.

The plan to get two rovers to the surface of Mars required that first we put together a package for Dan Goldin with a cover letter. Second, a 2003 task plan that had been agreed to by JPL and NASA HQ had to be prepared. There were other requirements: a second-rover budget plan signed off by the comptroller of NASA, a JPL internal review board statement recommending two rovers, an independent review board statement recommending two rovers, and finally a JPL Governing Program Management Council endorsement of all this. Dan may have tossed out his query almost on a whim and certainly made his decision quickly, but we had a lot of due diligence to accomplish, which the system absolutely demands.

We had to get a complete draft of the cover letter to Goldin done. We also had to get a signed copy of the task plan to JPL by 3 p.m. on Monday, August 7. We had to get a signed copy of the budget plan and get it faxed to me in Pasadena, where I happened to be at the time. We had to get a fax number for Dan Goldin, who happened to be in Morocco representing the administration at some event. We had to write a one-page summary of the JPL internal review board. We needed other items: a one-page summary of the independent review board by Gentry Lee plus a one-page endorsement of the board recommendations. We faxed all this stuff to Ed Weiler on Monday evening. We held a teleconference with Ed on Tuesday morning to get the official approval to proceed. We had to fax everything off to Dan Goldin that same Tuesday morning. Then, we waited—nervously. Dan did call me back from Morocco to ask a number of questions and said, "Sounds good. Go ahead." We notified our public affairs person to prepare a "note to editors" for a press conference that would be Thursday, August 10. Of course, we also had to notify the Office of Management and Budget, the Office of Science and Technology Policy, and the congressional science committee chairs. This had to be done on Wednesday so that we could have a Thursday press conference. There you have it: a typical week in Washington.

Ed Weiler gave me a copy of a letter from Ed Stone, the director of JPL. Weiler said I should put it in my files in case we ever went up for trial for this idea of having two rovers. The letter is very short. It says,

> Dear Ed,
>
> Based on all the study's findings and reviews, I am prepared to commit the laboratory to proceeding with the two rover option for the Mars 03 opportunity. Mission success will be defined as successful rover operations by at least one rover. Proceeding past phase C [a reference to NASA mission development stages that corresponds to the actual build of something] with the two rover option is contingent on developing an operations plan that minimizes conflicts between the two rovers and the other concurrent high-priority deep space missions. The Mars team was not able to demonstrate such a plan in the limited time available for their study. The two flight systems, including the payloads, will

> be identical. Final cost commitment will be made at the end of phase B, and FY 01 funding will arrive within the first week of the fiscal year. In the meantime, I assure you that we are enthusiastically committed to achieving a fully successful landing on Mars in 2004.

All things considered, I think Stone gets a lot of credit for making relatively few yet vitally important demands and finding a way to embrace what was essentially a sea change.

The material that we faxed to Dan Goldin on August 8 had all of the things that we promised. They had all been developed in a period of, perhaps, two weeks. Who says the federal government is inefficient?

The press conference went extremely well. We had our talking points. We had our overviews. We had our plans. Comments at the time were that we were playing a high-stakes game. The coda to this whole story is that after the press conference, we got a call from CBS's morning program, *The Early Show.* At the time, *The Early Show* was co-hosted by Bryant Gumbel. Gumbel was noted for his abruptness and rudeness to his guests. This personality trait is probably what got him fired from the show.

Originally, *The Early Show* wanted Goldin on, but he was on vacation, so I got the wonderful opportunity to be interviewed first thing in the morning. I remember distinctly that Gumbel's first words were, "So you fly two rovers to Mars. Why?" I told him that the science of Mars exploration was exciting, that we were following the water to find the fingerprints of life, and so on. He persisted, though—why two rovers? I confess that some part of me was tempted to blurt, "Because the big guy told me to!," but in the meantime, I had become convinced that not only was it not such a crazy decision, it actually was the right thing to do for scientific and programmatic and even engineering reasons. I explained that they gave us more science and that also meant that we had a much higher probability of one succeeding. Gumbel followed this track by saying something to the effect of "Why? Don't you think both are going to succeed?" To which I responded, "Yes, of course. That's our goal, but you never know what Mars will throw at you." Gumbel then grumbled that it seemed like we were wasting the taxpayers' money on two rovers and that he wasn't sure why NASA was doing this at all. I countered the best I could, saying that we felt that the science of finding out whether we were alone in the universe was extremely exciting to all individuals. Gumbel didn't really care for that answer, it seemed, but at the end of the interview, he relented and said that I should have a nice day and signed off.

Ten years later I think we have demonstrated that Gumbel was wrong in his assessment. We have used the American taxpayers' money extremely well, and learned a great deal about Mars. We've excited kids and students about science and exploration, and I think the two rovers have been as successful as the Hubble Space Telescope in engaging the American public in the excitement of space exploration.

CHAPTER SEVEN

Hands across the Water

The space business has been an inherently international endeavor since it began. In the earliest day, the Russians had a success, and we jumped in. We had a success or two, and they matched us and tried to best us. We struck back and so on, but space is a grand, risky, and very expensive arena in which to operate, and eventually competitors become collaborators. Today, we rely on the Russians for significant operations connected with the International Space Station, and we routinely collaborate with the European Space Agency (ESA), various specific European space agencies such as Centre National d'Études Spatiales (CNES, the French space agency) and Agenzia Spaziale Italiana (ASI, the Italian space agency), as well as many others like JAXA (Japan Aerospace Exploration Agency) and CSA (Canadian Space Agency). No one is truly able to go it alone in space for long, plus there are diplomatic advantages to playing nicely in space. We do all truly live in the solar system together.

The new Mars Program had a strong international component. The essence of this collaboration was around the Mars sample-return mission that had been planned for 2003 or 2005. I had never felt comfortable with the notion of what looked to me like a premature sample-return mission. I felt strongly that there were insurmountable issues on both the science and the technology sides. We at NASA HQ had canceled the U.S. planning for this mission, and we needed to set about organizing a new relationship with the Europeans. The thought of cancellation of the Mars sample-return mission was very unpopular with groups at both JPL and CNES. We were told time and again by people at the French space agency that they had raised their money based on a promise to carry out a return of a sample from Mars. We planned a trip to Europe in June that required us to be on our best diplomatic behavior.

In Washington, D.C., when you set about conducting international discussions and international collaborations, you don't do this on your own. You have "help" from people in the international affairs office of your agency, in this case NASA. You get help from the State Department. You get help, which in some cases is really valuable, from the American embassy located in the country to which you are traveling. One of the first things that happens is that you get issued a red passport. This is the official U.S. passport for government employees and at times is more a liability than an advantage. The red passport, if you were on a plane that was being hijacked, would identify you as a government employee and could put you at risk. So we were advised that sometimes it's better just to use your blue personal passport. Luckily, I found the people in the international affairs office at NASA HQ to be universally helpful. Folks like Diane Rauch, who helped organize a trip to Russia, and Steve Ballard, who helped us with the trip to Italy and France, were very knowledgeable and very helpful.

The e-mails, official letters, telephone conversations, and the like highlighted the need for the always crucial face-to-face meeting sooner rather than later. By June 2000 my team at HQ and Firouz Naderi and his team at JPL had a pretty good idea of what might be in the new Mars Program. We didn't have all the details filled in, we didn't have all the cost estimates finished, but a picture was emerging. And in that picture, which ultimately ended with a Mars sample return sometime in the next decade, there were obvious places for international collaboration.

At this point I have to acknowledge that working internationally is clearly a two-edged sword. First, there's the political dimension. My boss on a daily basis, Ed Weiler, was distinctly divided about the value of international cooperation. The White House administration at the time, under President Clinton, was very much in support of international collaboration, but over time administrations and their attitudes toward cooperation come and go. The Mars plans needed to last a decade. In addition, the people at the front lines of putting programs together had seen how complicated it can be to work across eight or nine time zones with several languages and multiple interfaces. And these are just some of the more obvious issues about international efforts. Their administrations and funding streams are subject to change over time too. Consequently, I went into the discussions with our two principal partners, ASI in Rome and CNES in Paris, with distinctly mixed emotions. The trip was finally set for June 11–16, 2000.

As is customary in these cases, the diplomatic groups—that is, the people in the international affairs offices—worked on an agenda, which I looked at and I'm sure my counterparts in Paris and Rome looked at as well. In the end we agreed that we would review all the things that had been discussed in the past, before I came on board, but that our overall goal would be to find a way to work together in what was emerging as the new plan.

Our first visit was to CNES. As I found out, it is customary for the international affairs people at NASA HQ to work with us managers and executives to develop a position so that we knew what we had to do before we ever arrived in Paris.

The initial step, though, was on this side of the pond. The principal French space agency people, Richard Bonneville and Christian Cazaux, visited us in Washington on May 17. Bonneville is very much a Parisian, a northern Frenchman, very intense, intellectual, carefully examining and scrutinizing all of our statements. Cazaux was much more relaxed, a southern Frenchman, very Mediterranean, very much a friendly guy to have dinner with and very easy to talk to. During the discussions in May the chief topic was, of course, the Mars sample return. We were reminded again and again by Bonneville in particular that their minister had raised several hundred million euros for the French and European space effort aimed directly at the sample return. We had to tell Bonneville and Cazaux now that we didn't see any way that we could do the mission as planned as quickly as had been originally discussed. This, of course, made them extremely unhappy, and in large part, I think, because they were fearful of losing the investment their minister had garnered. Nevertheless, the sample-return mission, in my view and the view of our kitchen cabinet, could not possibly be done on such a short schedule.

As a result of the discussions on May 17, we developed a position that we carried with us on the trip to Paris later in June. First, we intended to reaffirm our plan to form a long-term partnership with CNES in the area of Mars exploration. Second, we would reaffirm what the French space agency representative said was critical: that their role be significant, visible, and technologically challenging. One recurrent theme was that the space agencies in Europe clearly looked toward developing their own industry bases and their own capabilities as a result of collaborating with NASA. Third, we were to reiterate that the Mars Program was undergoing a major replanning that would not be completed until near the end of 2000. Until we'd finished that, we couldn't commit to any specific details. We emphasized that NASA would like CNES to recognize this schedule and be flexible. And we invited the French to fully participate in the preplanning and asked that they serve on our executive committee and jointly participate in a workshop in July, jointly organize synthesis activities, and help coordinate input from the French science community and industry.

We also concluded that we should welcome CNES participation in independent studies and independent program concepts like the ones they presented to us in Washington on May 17. We were obliged to tell the French that we were very sensitive to risk these days, after notable failures, and we would ask the French space agency to explore any mechanisms for risk management. One mission concept that CNES, and the European Space Agency, had brought up several times was called "Netlander." Netlander is the idea that I had worked on under the name of MESUR, the Mars Environmental Survey. The idea was to set a group of small landers on the surface of Mars for the purpose of making seismic measurements and therefore determine the inner structure of the planet.

After the flight over and a day of recuperation, on June 13 and 14 we sat around a table at the headquarters of CNES in Paris. Following the usual preliminaries and greetings, very important in Europe, I laid out the Mars exploration planning process that was under way and our objectives and our constraints. Then we had a long discussion of what the various options and common objectives might be. They took us to lunch at Au Chien Qui Fume, the Smoking Dog restaurant. I came to learn later that it was a tradition of the French space agency to take visiting dignitaries to this restaurant. It was within easy walking distance, and it was a typical French dining experience, I think. We had three courses and a dessert with multiple servings of wine and a champagne toast. After all of this, it's very hard to think about anything, let alone delicate diplomacy, and then we adjourned at about 4:30 to get ready for a big dinner that evening, another multicourse French meal.

The organizing principle the French space agency chose for its participation in Mars was called Premier. As with NASA, all major programs are given names, often bold or gallant or ambitious names, designed to inspire those who work on them. Premier, no doubt referencing the notions that the program was of the highest quality and that it represented a foray into a new realm of exploration for France, included two main trips to the International Space Station, a Mars sample-return mission with NASA, plus a Mars lander network in cooperation with European partners—decidedly an ambitious program.

Our CNES colleagues reminded us again, and again, of the importance of a Mars sample-return mission. We had to tell them again, and again, that we just didn't see any way this mission could be accomplished or the technological and scientific challenges developed on the time scale previously discussed. It was a very slow and painful process over the year and a half I was Mars program director to convince the French that a sample-return mission was simply not going to happen until the next decade. One of the tools that I employed to get them to understand both the scientific risks and the technological challenges was to involve their community and their leaders themselves in our replanning process. It was our thought that by having them sit there with us, hear all the presentations, and go through the deliberations they would come to understand how difficult this all was. In the end I think that Richard Bonneville and Christian Cazaux both came to understand that a sample-return mission was going to be deferred until we had answered some important scientific questions and developed a series of new technologies. There was, and continues to be, a very vocal member of the French science community who maintains that a sample-return mission could have been done in 2003 or 2005, should be done now, must be done ASAP. As far as I can tell, he was never convinced by any of his colleagues on either side of the Atlantic. Mars really is hard, and not just because of the science and engineering.

After all the discussions held between members of CNES and NASA, in the end there was actually very little collaboration between the two agencies in the final mission queue that we presented to the world in October 2000. I think this was mostly because we were not able to find a way to significantly advance the Mars sample return in the way that the French had hoped. I believed then, and I believe now, that understanding Mars as a system by carrying out the scientific reconnaissance necessary to pick a sample was the most important thing we could have done. The additional funding was simply not there to work on the technologies and the mission components that the French wanted to pursue and that sample return requires. Individual French scientists, of course, were co-investigators and members of various teams that have been part of essentially every mission that has been flown and are great advocates of the program, but it is very likely that a Mars sample-return mission will not commence in any form until close to the year 2020.

We next moved on to Italy, to Rome. The key individuals there were Enrico Flamini and Marcelo Coradini. Enrico and I had come to know each other before we ever started working together in my new role as Mars program director. Enrico's family history, he told me over dinner one night, went back some five hundred years in Italy. He had been working with the Italian space agency, ASI, for some time and had a PhD in physics; his dissertation was on x-ray analysis as applied to extraterrestrial materials. I found him to be a very sincere collaborator who was doing his very best to work within the constraints that we all have within government service. Enrico and I have worked together since that time, and now I can say I have had him as a friend for more than fifteen years.

The Italian government has changed perhaps fifty times since World War II. Although a lot of the individuals seem to stay in the government orbit, nevertheless each new government feels obligated to change what the previous group had committed to. The discussions that had been going on previously about the Mars sample-return mission and the change that I had put in place clearly meant that we had to do some negotiations, some fence mending, and some smoothing of ruffled feathers.

In preparation for this face-to-face meeting with ASI, we had a teleconference on May 22, which included not only me and some of my team in Washington, but also, in Rome, Enrico and his science leader, Marcelo Coradini. As a result of that teleconference, we developed a plan position before we ever got on the airplane. We would reaffirm our intention to form a long-term, significant partnership with ASI in the area of Mars exploration. We also had to underscore that our entire program was undergoing major replanning, which would not be complete until the end of 2000. We would relay that until that planning was completed, we could not commit to a specific implementation or a specific scenario. We would ask that ASI recognize the schedule and remain flexible as our new plan emerged. We would invite ASI to fully participate in this replanning in a significant fashion. We would invite Enrico to serve on our executive committee. We would suggest participation in a workshop we had already set in motion for July and in the synthesis activities to pull everything together. We would also ask Enrico to coordinate input from the Italian science community and their relevant industry.

We NASA folks welcomed the ASI initiative to come up with new ideas and generate program concepts, some of which they presented over the phone on May 22. We also had to tell the ASI people that NASA was very sensitive to risk these days, and we encouraged the Italians to explore risk management by strengthening the telecom infrastructure at Mars, the all-important telecommunications network that sends both science and engineering data back to Earth. This was our not-so-subtle way of suggesting to them that if they could come up with a way of helping communications with what we believed would be a large number of Mars missions, it would be well received. We finally told ASI that, although we would continue to try to advocate for substantial extra funding, we didn't see how the sample-return mission could be launched on the original schedule and that it would likely not be launched any earlier than the next decade. We asked ASI to factor that position into their planning.

In my presentation on June 15, my very first chart said in bold letters, "NASA wants to partner with ASI on Mars." Then we laid out the meeting objectives as we'd planned ahead of time and tried to focus on where we could work together. One of the things that we talked about numerous times was that we had to find a simple and straightforward way of working together. A complex interface would only raise the risk in our program.

As we had with the French space agency, we emphasized that international cooperation required that the partner bring a unique capability or expertise to the table. Our other criteria were that whatever was proposed really needed international participation, that it increased flight opportunities and science return, that it met the objectives that NASA had, and that there was value for a foreign partner. We defined examples of cooperation in the Mars Program, such as the Russian-provided neutron detector; the planned work with the European Space Agency on Mars Express, an orbiter that was ultimately launched in 2003; and some of the other concepts that had been put on the table.

In the end the most important thing that came out of the discussion with ASI was their commitment to providing a telecommunications orbiter. I confess that having seen the fluctuations in the Italian government over the years, I set aside, along with Firouz Naderi, a quite significant amount of funding to be sure that we had a communications orbiter team on the U.S. side as well. Sure enough, the communications orbiter eventually vanished as the Italian government changed several more times. However, one thing that Enrico and I did agree on was the addition of a shallow sounding radar instrument that would go on the next American orbiter, which was officially launched in 2005 and called the Mars Reconnaissance Orbiter. The radar instrument was called ShaRad, an acronym made out of "shallow radar" sounder. This instrument also turned out to be a very good complement to the deep sounding radar, Mars Advanced Radar for Subsurface and Ionosphere Sounding (MARSIS), aboard the European Mars Express mission. The data from ShaRad and MARSIS now show us what appear to be ice interfaces, buried glaciers below the surface of Mars that you can see with normal visible imaging.

I would end up spending a significant amount of time working with the Europeans to resolve our differences and to pull together a coalition that could bring the new Mars Program along in the decade of 2000 to 2009. I would make another trip to Europe the following year. Our invitations to the Europeans were accepted readily, and we worked together at the meetings that we had called in the United States. Over time we developed good personal relationships as well as extending the professional collaboration.

Similarly, the involvement of the international community had to be folded in with the development of a program that was primarily based in the United States with NASA. It is difficult to describe how all of these things occurred nearly simultaneously. While the storytelling is best followed in a linear fashion, the truth of the matter is that on any given day, I would find myself jumping back and forth between the 2001 launch preparation review, a 2003 definition review, international discussions about missions further in the future, and the development of advanced technology, plus probably have to attend a party that evening.

Because Washington, D.C., is the geopolitical center of the world, "every night is prom night," as my wife noted. I found it necessary and cost effective to buy my own tuxedo and did my best to convince Susan that soggy canapés and warm Chardonnay could be fun.

CHAPTER EIGHT

Getting the Word Out

If the story appears simple or straightforward in any way, don't be fooled. It wasn't. On the one hand, Firouz Naderi and I were in sync developing the program systems engineering approach, and the science community was embracing the elegance of "follow the water." However, never underestimate the ability of Washington politics and the aspirations of an optimistic and ambitious administrator to muddy waters that are threatening to appear crystal clear. Slogging through the Washington process is, by its very nature, repetitive, frustrating, often circular, and seemingly endless. There were times when I began to question if, having gotten so many things right, as far as I was concerned, we would ever be able to implement the plan to the satisfaction of the various bureaucrats involved. Still, the process—laborious, repetitive, and meeting intensive—is there when one is spending federal taxpayer dollars, and amazingly, in the end, it may actually have served its purpose.

All of my restructuring activity, including the team at JPL and my NASA HQ team, was ultimately aimed at obtaining approval from the administration for a new mission queue, a new technology investment, and, finally, a new budget to make all that a reality. Getting the agreement of the key gatekeeper, Steve Isakowitz, at the Office of Management and Budget (OMB), was not an overnight event. As soon as I was named program director, there were e-mail discussions between Ed Weiler and Isakowitz. My first face-to-face encounter with Steve Isakowitz and his lieutenants, Brant Sponberg and Doug Comstock, came on May 12, 2000. After that meeting I received a long list of questions and examples from Sponberg of what they would like us to produce. These included things like, "What are the compelling science goals? What is the integrated mission set? How do you distribute risk across multiple platforms?," and on and on. Isakowitz and Sponberg noted that we needed to embrace the 1996 national space policy goals of a sustained program to support a robotic presence on the surface of Mars. Sponberg also had a large number of questions about what kind of technology investments would be required and how to measure whether or not you're successful. He concluded his list of interrogations with a comment about making sure that whatever we did would engage the public and educators through direct and interactive means. Because OMB is the ultimate gatekeeper for putting budgets in and submitting them with the rest of the president's budget to Congress, its opinions are very important. There was clearly a lot to juggle, but actually I was relieved to find that their questions and concerns were good ones and reflected a lot of the thinking we had already been doing.

The timing of all this work was fast and furious, and various activities were interwoven with others. None of us had the luxury of uninterrupted pursuit of carefully planned actions. The JPL meeting in early May was followed by an international trip, then the decision meeting on the 2003 mission. That quickly morphed, as seen in chapter 6, into the two-rover decision.

My first synthesis meeting over August 21–25, where we attempted to literally bring it all together, was followed by a second synthesis meeting September 6–8. By that time, early September, we had entered the review season. I had no fewer than three reviews of the new Mars Program with Dan Goldin, and a full-up meeting with the Tom Young committee on October 10. Each of those meetings added some bit of clarity and focus to the plan that was emerging.

The number of stakeholders was truly impressive. There was seemingly, at times, an endless list of people involved, all of whom had opinions and all of whom could affect the outcome in one way or another. We had reached the point in the process where we were briefing individuals, small groups, and some larger groups, and, of course, that meant PowerPoint charts. All jokes about PowerPoint aside, it is a clearly convenient tool and maybe even a necessary evil when complex ideas have to be communicated in finite periods of time. We needed buy-in and decisions, and we needed them rapidly. We had to have clear communication tools. It is a truism in Washington that there are many people who can say no, but very few people who will say yes.

The second critical OMB meeting was in July 2000. Dan Goldin, of course, kept constant track of how our planning was proceeding because the failure of "faster, better, cheaper" was a blow to his overall philosophy, and the international visibility of the loss of the two Mars missions made successfully restructuring the Mars Program a priority on par with the International Space Station. Goldin really wanted to redeem both himself and the agency. Audacious thinking had led to the two-rover concept, so soon we had a proposal to go to the OMB right after the Fourth of July with an audacious and far bolder Mars Program than we had been considering.

The suggestion was made that the program take the baseline that we had so far—the "ladder to Mars"—and look at the possibility of adding an orbiter in 2003, come up with a robust $100-million-a-year technology program, and then request the funds to have what was called an "aggressive Mars exploration program" with additional missions, orbital reconnaissance, sample returns, data analysis, a communications network, and so on. When I added all of this up it amounted to a program that was at $1 billion a year, far greater than the current budget. Still, when your boss's boss says jump, you'd better at least attempt to respond with a bit of air, so with Goldin's blessings and encouragement, I took Jim Garvin over to see the gatekeepers, Steve Isakowitz, Brant Sponberg, and Doug Comstock.

Our reception was less than enthusiastic, although as I came to learn over time, Steve was, with his staff, a genuine fan of Mars exploration. Still, the idea of increasing the funding for an expanding program from $350 million to $400 million a year to over $1 billion a year was not credible. Steve asked many, many questions about exactly how we would spend this money, what constituted an aggressive exploration program, how we would spend so much money so rapidly, and then sent us back to the showers to come up with a decidedly more realistic program. This would not be the last time that Dan Goldin and his enthusiasm would cause me and our Mars program staff to generate a whole series of options that had only a limited chance of success in passing even a preliminary gate.

As this process of developing a new program evolved, I became aware that I needed two separate efforts at the same time: One was being responsive to Goldin and ideas about what the new program might be, ranging from the brilliance of two rovers to the grandiosity of accelerating sample returns, creating communications networks, and additional poorly thought-out bells and whistles. The second effort was the one I had begun as soon as I had agreed to serve as Mars program director, and it was much more deliberate: Canvass the community; look at options; assume that while we might get some extra budget, it would not grow enormously. This effort was the one that I pushed until we were able to announce publicly what our plan was for the decade.

By the time of the second synthesis meeting in early September, we had developed three options to address my dual task of serving both the administrator and the broader community. These were the options I carried to the first briefing with Dan Goldin on September 14, just a week or so after that synthesis meeting. Of these options, the first was called the "most desirable," despite being somewhat out of the resource (budget) box. It represented an aggressive approach. This was, after all, what Dan had requested. It met all the high-priority attributes of robust science return and included a Mars sample return before the end of the decade—by 2010. It had resiliency, separate mission paths, and a strong technology program. It also relied on significant early international collaboration.

This most desirable program started with the assumption that we would launch an orbiter in 2001, the twin rovers in 2003, and then in 2005 we would have a more capable rover launched by the United States with a variety of payloads (science instruments, essentially). Next, the Italians would launch an orbiter with a payload that was being developed under the name of Science Orbiter, and also in 2005 the French space agency, CNES, would launch an orbiter with the U.S. contribution of a payload capable of rendezvous and returning Mars samples. In 2007, we would complete a principal-investigator-based Mars mission. This notion of competing for creative missions, rather than determining defined missions far in advance, is what eventually evolved into the Mars Scout Program and turned out, like the twin rover notion, to be a real asset to the final program. In 2009, there would be another orbiter provided by the French with the U.S. Earth entry vehicle and rendezvous payload on a U.S. lander. That same launch would have the Mars ascent vehicle and other things needed for sample return, and the Italians would provide a telecommunications orbiter on a separate launch. In 2012 we would have another competed Scout mission. Phew.

The second queue option was one that was within the resource box we expected to have, and this mission set assumed that international participation would be minimal. Privately, I had begun to conclude this was the likeliest and best outcome, as it was by far more predictable and therefore less risky, but much more drama was yet to unfold. This option delayed a sample-return mission until after 2010. For this option, there was, again, the Mars Science Orbiter in 2005 with the reconnaissance payload under discussion. This would be followed by a competed mission in 2007 and a U.S. smart lander in 2009. The lander was at that time dubbed the Range Rover, and it would demonstrate advanced entry, descent, and landing

(EDL) and be something similar to what would be used later for a sample return. In 2012 we would begin the sample return with the orbiter, followed by the lander in 2014.

Option three was also in the cost box but assumed that there would be significant international participation that would lead to a sample return before 2010. We had to leave such participation on the table, despite the inherent schedule and therefore cost risks, primarily for political reasons. In this program the United States would launch a lander in 2005, but it would be a stationary lander—the lowest cost possible to demonstrate advanced interest in landing that would lead to a sample return–rather than a rover. In 2007 there would be an Italian orbiter with a Mars Science Orbiter payload that would share a ride on an Ariane 5 (a French launch vehicle) with a French orbiter. The French orbiter would carry some surface stations with a U.S. contribution of the rendezvous and Earth return vehicles that are necessary to make a sample return possible. In 2009 there would be a French orbiter with an Earth entry vehicle provided by the United States. Once again, it would share a ride on a U.S. vehicle and the U.S. lander with the Mars ascent vehicle that would also be launched in 2009. We had the potential for an Italian communications orbiter, followed in 2012 by a competed Mars Scout mission. We were clearly trying every way we could to maintain the French engagement and their promised investment of several hundred million euros.

If these various options sound arbitrary, repetitive, and complicated, in many respects they are. From outside the system, it surely appears wasteful to put serious effort into three separate threads of planning, and the differences might even seem less than compelling from the outside. To us, though, we cared passionately not only about accomplishing the task at hand as well as we could, but also about the exploration itself. Added to that, of course, for the various groups with money, jobs, pride, and their own passionate scientific curiosity at stake, every element of the final queue had a significant impact. Part of my job, very explicitly, included “responding to stakeholders,” the fool’s errand of attempting to be all things to all people at all times. The minefield of competing needs and desires is precisely why we wanted to be so methodical and detailed in our basic approach. Whatever we did, some folks were going to be unhappy. We had to try to minimize that and at the same time arrive at a real-world solution to our task that had the greatest potential for mission success. After all, we really were talking about sending lots of sophisticated hardware and software to a distant planet. We would be airing our decisions publicly. The August 10 press conference had been solely to announce the two rovers. That was a done deal. In October we needed to face the world with a solid, stable, achievable ten-year plan.

It was clear from the moment that I accepted the job until we finally presented to the public on October 26 that Mars sample return and how to implement it dominated the discussion of the program. The Mars sample-return mission was, and remains, a very high scientific priority; however, it is also expensive and technologically demanding. Before we went to see Dan Goldin for the first briefing on September 14, we had our own discussion internally about how to deal with Mars sample return. First, we grappled with "Do it this decade," that is, by 2010, or defer it until later. Many scientists, especially from the astrobiology community, believed that we needed to understand the surface of Mars much better through reconnaissance before we selected the perfect site or even a compelling site for sample return, and if this took time, so be it. Then there were those who believed that by late in the decade of 2010 we would've learned enough to pick a reasonable site, and we should simply go for it. Some of these members of the science community advocated what became termed a "grab sample." That is, they argued that the first sample from nearly any site, as long as it included rock fragments, would give us tremendous knowledge about Mars. Let me point out here that the sample was likely, no matter what, to be tiny—a few grams of soil with some pebbles mixed in. It can hardly be said often enough that Mars is hard.

Our international partners, particularly the French, conditioned their participation on sample return by 2007 or, at any rate, no later than 2009. The reason for this was the agreement they had with their minister of science, who had personally obtained a commitment for sample-return funds. If the joint French/United States program could not move quickly enough, those funds in France would be used for other things, and the commitment would expire.

I had already taken the position at the first synthesis meeting in late August that a very compelling science argument was needed to pursue a $2 billion sample-return mission (in year 2000 dollars). In addition, we had identified a long list of advanced technologies that needed to be developed before a sample-return mission could be pursued. Those technologies included the Mars rendezvous in orbit, precision landing, hazard avoidance, long-range mobility, sample transfer, the Mars ascent vehicle, automated deep space rendezvous, and all the constraints associated with being sure that we neither contaminated the surface of Mars nor allowed any return sample to contaminate Earth. The requirements list was formidable, and I was well aware of the fact.

Thus on September 14, 2000, we walked into the administrator's conference room on the ninth floor of the NASA HQ building. Long and narrow, the room was dominated by the table that nearly filled the room, forcing late arrivals to slither sideways to get to the chairs at the end. I was armed with a mere ten charts for what I expected to be a brief initial meeting. We had two possible paths to meet the demand for a sample-return mission. One was to break up the technologies and the missions into pieces and demonstrate them in various precursor projects in what we call "feed-forward" technologies. In that scenario, which is close to what we have been doing the last ten years, we would have completed the precursor missions by the time of the detailed design review of the sample-return mission in order to get the results fully integrated into the development path. The alternative was to do what Viking had done: Build two very large, multiple-spacecraft projects and launch them both. This required substantially more budget than was available.

On September 25 we had a second, more formal briefing with Dan Goldin. During the first briefing, on September 14, Goldin had asked a long list of questions about risk, cost, and international collaboration. He had wanted to know about the reconnaissance we would carry out for the next generation of missions, how we would include small missions, and, ultimately, what we were going to do about sample return. So for this second meeting we had fifty charts and told him that for five months, we had gathered data and performed analysis sufficient, we thought, to define a new program. We believed, though, that a year's worth of program systems engineering would be needed to reduce the uncertainties connected with sample return. We pointed out that whenever you do the sample return, that process will dominate the program for three full launch opportunities, or at least six years. As to a human exploration component, we thought we had included that, to the degree possible, within the resource constraints we had. Also, we had to fight to budget needed technology investments.

It was at this second meeting that we told Dan Goldin that Tom Young would review us on October 10, that we would probably have one more meeting with him, Goldin, and that we would like to announce the plan, if possible, *before* the next meeting of the Space Science Advisory Committee, November 1–3. That meant that our window for announcement was somewhere around late October. The French and Italians were proposing major program elements beyond just science payloads, and if they were going to be included we would have to have letters of intent beforehand (standard NASA practice). Spain, through Juan Perez-Mercader, one of my astrobiology colleagues, was proposing a modest communications element as well. We needed decisions in order to respond to the various external pressures. After discussion of a lot of technical issues, we presented Dan Goldin with two options, winnowed down and coalesced from our original three. One we called the "right program," that is, if money were no object the kind of program that you would put together. The second option we presented was one that was, within the uncertainties of the kind of estimates one is forced to make about one-of-a-kind complex efforts, inside the cost box.

As was customary in presenting to Dan, the conversation took abrupt turns and often paused for monologues from Dan himself. You always had to be prepared to jump right to your bottom line at any point in order to get a decision. The package we put together spent a lot of time describing the complex EDL issues associated with Mars sample return. We presented a decision tree of how you would decide which EDL approach was the best. We went into great detail on things like hazard avoidance, terrain sensing, and the various phases of atmospheric entry. We pointed out concepts for hard landers, semihard landers, and soft landers, such as those in the Viking style with retro rockets. Finally, we showed the evolution of rovers that we were anticipating and were quite deliberately planning.

We had a colorful chart that showed the much-loved and very successful 1997 *Sojourner* rover. It also showed one of the Mars Exploration Rovers (MERs) slated for 2003, and it showed, for the first time, the rover planned for 2007 that would be almost a metric ton (2,200 pounds) with tens of kilometers (10 to 20 miles) of range, using a radioisotope power source. This latter design should be successfully landing on Mars in the summer of 2012.

In large part responding to questions from our previous presentation to Goldin, we spent a lot of time discussing things like sample collection and how to provide planetary protection for both forward and backward contamination, and then moved on to the key elements of the new program. We listed the potential major international contributions, noting that all of them had questions of commitment, and in the case of the French, a very definite expiration date for the money.

By the time of this briefing on September 25, we had settled into a good high-level design for what we were calling the Mars Reconnaissance Orbiter (MRO). This mission would allow us to recover investigations that were lost when Mars Climate Orbiter disappeared, and it would provide us with high-resolution new data from the visible to near infrared imaging spectrometer plus very-high-resolution visible imaging. At the time, we were forecasting the possibility of 20-centimeter (8-inch) resolution on the surface. It eventually became about 50 centimeters (about 20 inches), but such resolutions have allowed leaps in understanding of Mars. We would essentially place a spy satellite around the red planet.

The small, competed missions from the Mars Scout Program were becoming much better defined. We thought that for the amount of money available, around $300 million at the time, platforms might include small landers, small rovers, gliders, or even airplanes or balloons.

We also emphasized to Goldin the absolute necessity of providing ongoing Mars communications and navigation. As the queue of missions began to be ever better defined and developed, we saw that large amounts of data would be returned and that our armada of spacecraft in orbit should provide the ability to relay data from the surface as well as help future spacecraft land more accurately. We had begun to understand that we needed either a dedicated communications satellite or at least a dedicated communications package on every orbiter.

So in our effort to get a decision, we presented two options. The first was the "right program" requiring almost $2 billion extra over the decade of 2001 to 2012 that would, however, provide dual sample-return missions within the decade, two telecom orbiters, and two competed small missions, plus significant technology investment as well as measurements for human exploration. This high-ticket option assumed significant international collaboration. The second option deferred the Mars sample return and defined a complete program that was in the current budget box, assuming the president's 2001 budget that was then being debated in Congress. It did not require any international participation and would provide advanced reconnaissance, two competed small missions, and a sophisticated lander that would prepare for sample return in the following decade.

Goldin, as usual, started raising questions, including a suggestion that we may have missed a bet by not looking at a Russian airbag system from forty years ago. Okay. It seemed that our much-needed decision was to be delayed a bit.

On September 29, we presented for the third time an assessment of the Mars Program to Dan Goldin. This was the last presentation before we took the whole story to OMB to get their blessing before we went public. Responding to Goldin's questions from the previous briefing, we described the work by an inter-center team on EDL. We answered questions about the difference between ballistic missile guidance and what is necessary in a much more shallow entry into the atmosphere required for a Mars landing. We also answered some very detailed questions about how many different events had been required for the successful landing of the 1997

Mars Pathfinder mission. The answer was fifty-two events, virtually all of which were firings of pyrotechnic (small explosive) devices that were redundant and of very high reliability. We pointed out that the Russian airbag system proposed many years ago had almost zero resistance to tears that might come from rocks and that the Mars Pathfinder system that we were proposing to reuse for *Spirit* and *Opportunity* was already designed and flight proven. This briefing repeated the planned evolution to a long-life, long-range rover, which seemed to meet with Goldin's approval.

We also addressed the question of risk for the 2005 Mars Reconnaissance Orbiter. Goldin had asked whether or not there was a significant difference between a single launch of the orbiter and two launches. By the time we got done pointing out that the funds required were an additional $240 million and that you would end up with two *Odyssey*-class orbiters rather than one much more capable spacecraft, Dan abandoned the idea of splitting up the mission. Those clear PowerPoint charts had rarely stood us in better stead.

Finally, after addressing all of the technical issues that had been raised by the administrator, we got down to discussing the budget options and which policy we would pursue. In the previous meeting, Goldin clearly wanted us to try again to maintain the European connection with the link to sample return. It was also by this point that my science partner in the definition of the new Mars Program, Jim Garvin, had come up with his "Seek/*In Situ*/Sample" mantra. Jim's presentation materials often featured triangles and this was no exception. Jim loved to show how he would go from doing reconnaissance, what he called "seeking," to landing on the surface and conducting *in situ* investigations, and then ultimately getting to a sample-return area. This thread was one that, for the science community, provided a context for the mission definition.

I presented, once again, two options, one of which involved significant international participation and one that fit in the cost box. I detailed various missions and carefully pointed out pros and cons for each. I included plans on the drawing board for a second Mars sample-return mission in 2013/2014. We pointed out that there were various options that could be calculated and taken as a de-scope (to redesign a project because of budget cuts) at the program level, but there were risk issues in having only a single lander, for example, as part of a sample-return mission. I was beginning to feel like I had been sucked into a nightmare version of Risk or Stratego or some other of those board games from my childhood where we moved cool game pieces around a board trying to feel powerful and grown up. We ended up with no clear decision.

It was after this third meeting with Goldin that I pleaded with Ed Weiler not to have any more meetings with Dan, lest Dan make some other suggestion that would lead us down a rabbit hole into more effort that was a dead-end proposition. I was pretty well convinced that we had all the elements we needed. It was now just a question of budget and degree of international cooperation. In fact, to avoid seeing Dan Goldin and getting some unanticipated direction from him, I took to going up and down the fire escape steps and using the freight elevator in the back of the building. I would joke at this time about glowing in the dark when I got home in the evening after having so many G (Goldin) doses and receiving this many or that many G-rays from the administrator. I know that Goldin's desire was to create a great Mars Program, but he had difficulty staying out of the details of the technical description and unwittingly asked us to do enormous amounts of additional work in what mostly became blind alleys.

With the third briefing to Dan Goldin completed, we were ready to see the Office of Management and Budget. Steve Isakowitz, Brant Sponberg, and Doug Comstock were all there at the October 6 meeting, which was originally scheduled for forty-five minutes and went on for almost three hours. The first thing Jim Garvin and I did, as always, was to describe the central characteristics of the program definition, that it was science driven and to be carried out using robotic spacecraft. We also said that we were still intending to provide information for the eventual human exploration of Mars. I covered all the outreach and data gathering that we'd done, including requests for information from industry, a workshop at the Lunar and Planetary Institute in Houston, concepts from various NASA centers, responses from the international community, and others. I described briefly our synthesis retreats, the makeup of our international executive planning team, and the program reviews that were yet to come. Those included, of course, the Tom Young review on October 10 and our desire to have a public announcement in late October that would be followed by meetings with the public advisory committees.

Along with describing program systems engineering, I repeated much of what had been told to Dan Goldin about the big Mars sample return. I indicated that it dominated the discussion, and I realized it was of unequivocal high priority, but it also had a number of technology challenges in addition to the science issue of sample selection. I pointed out that international participation could be a major contribution, but that the two principal participants, the French and Italian space agencies, would have to adjust to a long-term program rather than a one-shot mission or a single effort like the sample return. I reported that the community had rallied around the need for a Mars competitive opportunity in the small Mars Scout missions. I included the "ladder to Mars," and I emphasized that all the mission queues that we were going to present did, in fact, follow water and prepare for ultimate human exploration.

At this point I turned the presentation over to Jim Garvin. He used his triangular charts to couple life, climate, and geology, to point out the differences in the progression from seeking (reconnaissance) to *in situ* (surface investigation) and ultimately to sample return. Steve, Brant, and Doug all had a considerable number of questions about our programmatic preparation as well as the scientific work that had been done. As they quizzed us about reconnaissance, Jim pointed out that, over the course of the next decade, we would be going from thousands of potential sites for a sample return to hundreds, then tens, and then a very few sites that would be rich enough for a sample-return program.

We finally reached the meat of the discussion. With no clear direction from the administrator, we had retreated to three options: a robust, comprehensive program that required almost $2 billion extra and included substantial involvement of our international partners, a baseline program that needed some extra money but also had substantial international collaboration, and within the existing budget a U.S. program only. The three options were very similar to those I had presented just a week or so before to Dan Goldin. I emphasized all the constraints in dealing with international partners and that the French, in particular, demanded that a sample-return mission take place by no later than 2009.

Each of the mission components was then examined in detail by Steve and his team. The 2005 Mars Reconnaissance Orbiter was fully justified because it would recapture lost science from the previous failures and return extraordinary findings with new very-high-resolution visible and infrared imaging. The competed Scout missions and the attendant possibilities were very well received by Steve and his team. The notion of providing an opportunity for new ideas under a cost cap was as appealing then as it is now. It was when we got to the planned 2007 "smart lander" that the questions became more penetrating. "What would be the contribution of this lander to future sample-return missions?" they asked. "What would be the kind of science that might be done?" they questioned. (Note that the actual launch date was later changed to 2009 for technical reasons.)

My response to the contribution to future sample-return missions was very straightforward, as we had been talking about the need for precision landing and long-range roving sample acquisition and so forth for nearly six months. The science that could be done there was a question that had not been fully resolved. We agreed on the principles of the Mars Program, the "follow the water" theme, and the idea that among all of the priorities, the first among equals was the search for life. Over the next two weeks we refined the message and had numerous iterations via e-mail and telephone with OMB. It was necessary that we conduct most of this work at NASA HQ. I kept JPL and Firouz Naderi informed, but the rules of engagement are that HQ communicates with OMB and Capitol Hill; the centers talk to HQ.

The feedback that we got from OMB was the following: The public release should avoid going into detail on the Mars Program that exceeds the president's budget. Each of our stakeholders had a specialized perspective. Dan Goldin wanted the most spectacular program he could get for his agency. The OMB was the watchdog for budgets, and they wanted things to be fiscally responsible. Agencies shouldn't be promoting activities above the president's budget plans and a higher-priced option could make the existing "in-guideline" option look less attractive. Therefore, if there wasn't more money forthcoming, it could undermine support for it. Next OMB told us that we should be prepared to answer questions about the increase in cost over the previous architecture. I agreed some of the cost increases were differences in how we book the missions and some of them were actual increases in reserves and in cost of hardware.

OMB suggested not overemphasizing the importance of sample return because it could undermine discussions for the fiscal 2002 budget that was under consideration. It could also make the existing "in-guideline" program look weak because we had bumped sample return at least six or eight years forward from previous discussions. OMB felt that the discussions they'd had with us had shown that there were a lot of very exciting and very important missions that would precede sample return. They told us not to place "prepare for human exploration of Mars" at the same level as the search for life. Remember, this was in the year 2000 just before the presidential election between Al Gore and George Bush. For the then Clinton administration, human exploration was frosting on the cake, not the cake. Therefore, we should suggest human exploration along the lines of "Well, by the way, for the same price as search for life you also get information about looking at future exploration with humans."

On the issue of risk, we had made the statement to OMB in previous briefings that failures were not acceptable. OMB recognized that if you commit to not having any failures, that can drive up the cost significantly. On the other hand, taking too much risk, as had been done before with the resultant loss of not only an orbiter and lander but two small experimental probes, demonstrated that a shutout was unacceptable as well, so they urged us to be as robust as we could within the constraints of the budget.

There was quite a lot of discussion about what the compelling end state would be for the last mission in the queue. OMB advised us not to be so focused on sample return because the decision makers would view sample return as another Galileo (1989 launched to Jupiter) or Cassini (1997 launched to Saturn) flagship undertaking at a price tag of $2 billion per sample return. At that level, decision makers might question the value of doing it in the first place, thereby jeopardizing the entire option.

And finally OMB noted that total program results, not individual missions, would drive decisions over one option or another. At this point, OMB seemed to be asking that all the options should be presented and differentiated. This notation, from Brant Sponberg, concluded with a table containing five columns and fourteen rows that sought highly detailed engineering and scientific data. My notes from October 13, when I received the e-mail, were that this was far too detailed a table for OMB or the public, that it would change many times over the next year or so, and that it would be grossly misunderstood.

After the October 6 review with OMB, it still wasn't clear exactly which option we would be presenting publicly. We appeared to be getting, if not contradictory guidance, very broad and somewhat fuzzy advice. In the review with Tom Young and his committee on October 10 we covered a great deal of territory. The briefing package itself to Young's committee included several hundred charts and had to deal with things like breakdowns in the management system at JPL and technical issues like entry, descent, and landing, as well as the overall program composition.

What we presented to the Young committee is that we were proceeding with three options that varied in total budget from "in the box" to a plan that required almost $2 billion extra over the next twelve years. We also noted that the options varied somewhat from our consensus meetings and that of course the next administration would have a major influence as well. The good news was that all scenarios had the Mars Reconnaissance Orbiter in common. We would need to do an eighteen-month program systems engineering study for the elements after 2005 in order to get a proper program in place. A much more sophisticated long-life, long-range rover was common to all scenarios in 2007. Because of all the constraints on resources, launch opportunities, preparation of technologies, and so forth, the first sample return could occur no earlier than 2009 and might be as late as 2014.

The next sixteen days were filled with a flurry of activity. On October

12, I noted that we had passed the Tom Young gate and that October 26 was the most probable date for a press conference. However, our international partners were not yet on board with the program and the administrator was suggesting carrying three options. My own thoughts were that the program outlines were pretty well understood, but the details of 2007 and beyond were very difficult to sort out without making some clear-cut decisions. I spoke by phone with the Italian representative several times. He was pushing us still to negotiate a space on the 2005 mission for an Italian instrument in return for the proposed provided telecommunications orbiter in the 2007 opportunity. The French were very unhappy because the baseline option pushed sample return to 2009 or later. Their contractor was working on the old baseline. In Europe, space agencies don't have centers the way NASA does, and therefore they rely on fixed-price contracts with their industrial companies. It makes it very difficult for them to quickly reorganize their directions into something new.

On October 15, I sent a note to Firouz Naderi saying that our international partners were not in the bag, but I felt optimistic that we could come to some understanding, and if we couldn't we would have to modify our stories. We were still in discussions. The 2007 lander was particularly vague, and we needed to crisp up that story so that we would be clear what we were committing to. I included a request for the pictures and animation we needed in order to conduct a news conference.

On October 19, I discussed the full range of possibilities with Ed Weiler and Earle Huckins and noted that before we had the rollout, we had to be sure that all the different groups were on board with what we were about to say. Ed and Earle both told me to be sure that Steve Isakowitz was happy with our final plan.

A major announcement in Washington, D.C., from NASA requires a host of things that you don't necessarily see as you're watching it on television. There's a "note to editors" sent out a week to several days in advance telling them who will be speaking and when and where and what the subject is. The fancy animation, videos, photographs, and drawings that are so much associated with NASA press conferences have to be prepared. The press release itself not only has to be written, but needs to be reviewed by those who will be in the press conference: the bosses, the people in the public affairs office, the legal department, and, in this case, the Office of Management and Budget.

Amazingly, we were still keeping multiple options in play. I had prepared several versions of the "ladder to Mars," one with dates on it, one without dates, one showing considerable European involvement, and one not showing any international involvement. The "note to editors" went out October 24 announcing that we would have a press conference at 1 p.m. eastern daylight time on Thursday, October 26, 2000. The information on the new program would be the result of six months of intensive planning, involving NASA and international partners and building on the already announced plans for an orbiter to be launched in 2001 and twin rovers in 2003. The four people on the panel (four being the ideal number) were Ed Weiler, Jim Garvin, Firouz Naderi, and myself. Four is the ideal number because if there are more than four, either the individual talks are too long and you run over an hour or the talks are too short and you leave out information. Four is optimal for balance so that you can have two from the programmatic side and two from the science side to cover the full range of options.

Another media note: Press conferences are always held between 10 a.m. and 1 p.m. on a Tuesday, Wednesday, or Thursday. The time is chosen so that the media can meet their deadlines for the evening news or the next print edition. The days are selected so that the news will appear on a day when the public is engaged. Having a press conference on a Friday is known as "taking it out with the trash." Viewership and readership are very low on Friday evenings and Saturday mornings. Monday press conferences are avoided because the press doesn't like to travel on a Sunday and Monday evening has low TV viewership (except for *Monday Night Football*).

For a very important press conference like this you also prepare "dirty questions" and "frequently asked questions," so that you can anticipate what someone from the media might ask. Our public affairs people prepared a response to questions about the use of radioisotope power, which is always a very sensitive item because the public associates it with nuclear reactors and nuclear bombs, always a delicate topic even if the physics isn't correct. Thay's why launching any radioisotope requires White House approval. Under the dirty questions category were things like "How much is this program costing, and is it worth it?" and "What's the new president going to say about such an expensive program?" and "How can you commit to a new program when you don't know if you can pay for it?" Others we anticipated were "JPL was planning to do a sample-return mission in 2003 or 2005 for a much smaller cost. What happened to the plans for that mission?" and "What happened to sending two spacecraft every two years? Why have you decided to pull back?" as well as "Is this program really driven by science or does it happen to squeeze in science when you're testing out high-tech innovations?" All told, the public affairs people prepared twenty-three questions for me and the others to think through, so that we hoped we had a well-understood and well-reasoned response to what could be the media's inquiries.

The final press conference needed the approval of OMB. One of the really contentious issues was to what extent we would commit to international participation and to what extent we would actually set dates for a sample-return mission. We had prepared several different alternatives, including one that mentioned the possibility of human exploration of Mars in the future. However, it was literally not until about ten minutes before airtime that we received the message from OMB that we could only mention sample return as being in the next decade. We could not show a U.S. program that was dependent on an international partner, and we could not mention human exploration at all. Finally, we had some decisions. And they turned out to be consistent with what I thought was the most likely option among the two or three that we had been considering.

After all the comments, all the discussions, and all the preparation, my actual presentation had an amazingly clear focus. I began by saying that I was pleased to announce results of six months of intensive planning and restructuring. I noted that a significant effort had been made to address the management issues identified by the Tom Young committee. I was happy to say that we had now an excellent team in place, including my colleagues Jim Garvin and Firouz Naderi. We had new processes, clear lines of accountability, and more effective means of oversight and risk management. In truth, it was easy to be positive about all these elements, as I was genuinely proud of what we had achieved.

I went on to note that we were beginning a Mars campaign unparalleled in the history of space science. This process engaged an international Mars community consisting of several hundred scientists and engineers. In developing this plan, we had employed a three-part strategy of science, technology, and management. Mars Global Surveyor had taught us humility: Mars is very different than we had thought—it is not just a bone-dry desert. Garvin would give more details later, but in broad terms, our scientific strategy was to conduct an intensive phase of reconnaissance of the planet, which would provide an understanding of the possibilities for past or present life. We had adopted an approach that followed the trail of ancient and possibly modern water. I laid out three phases of technology development. I described a management goal that balanced scientific objectives, technology readiness, and budget to create a flexible program responsive to new findings and resilient to problems. I showed how a program is more than a mere collection of projects and how we would employ a two-track mission approach to give us time to respond to discovery and break the lockstep sequence of the past. We would seek before we sampled. We had orbiters for reconnaissance four years apart and landers for studying the surface with another four-year separation. It all led to a sample return. I talked about the specific missions and the great potential of the Mars Scout line of missions. When you believe in what you are saying, it's almost hard not to be eloquent.

Lest you get the impression that once decisions are made and budgets agreed to, you are out of the woods, I want to provide just one example of the myriad things that crop up in our business and help explain why so many space projects and programs seem to overrun their initial budgets or why they seem simply too expensive.

The visible, exciting, and publicly engaging part of any planetary program, or indeed any space program, is the missions themselves. Whether it's the science to be performed, the exploration to be done, the people who are involved, or the launch or the landing, they all excite the public, and they're all critical to a successful project. However, there is a part of the puzzle that, while extremely important, is not very visible to the public and not very sexy. That is the communications network, the collection of people, antenna dishes, and operations centers that transmit commands to spacecraft and bring the mission data, including all the wonderful images, back to Earth. For Mars, part of the communications usually has to do with orbiters around the red planet that can pick up data sent up from landers and relay those data back to Earth. Landed assets are usually tightly constrained, or limited, in their capabilities due to the complexity and high cost of landing anything on a distant planet. Since it's a little easier to get an orbiter there, the orbiters usually have the more powerful receivers and transmitters. They also have a straight shot back to Earth in part of their orbit. On the receiving end on Earth, NASA, as well as many of the other space agencies on Earth, uses something called the Deep Space Network (DSN). The DSN has three primary locations: in Madrid, Spain; in Goldstone, which is a town in Southern California out in the desert; and in Tidbinbilla, near Canberra, Australia. These three large antennas are positioned such that at any time at least one of them is in view of any given deployed spacecraft.

I had been briefed by JPL in early May about the communications network as an overall element of the program, and there was no sharply defined problem—then. Discussions by the previous program had been carried out and analysis had been done on creating some so-called micro-mission communications orbiters that would be at Mars and capable of sending data back. These micro-mission concepts, once they were scrubbed and looked at in detail, didn't really hold up. Their designs were what we call single string—that is, with no backup or redundancy. The cost estimates didn't seem to be realistic, and the definition didn't seem to really fit in with the overall architecture. We didn't know at that time in early May that we would have to incorporate some type of communications capability so that the new instruments we were developing and the new missions we were creating would be able to get their very large amounts of data back to Earth. For example, older missions such as Pioneer 10 and 11 (begun in the early 1970s) and Voyager (launched in 1977) were generating a few thousand bits per second or, in the case of one of the Pioneer spacecraft, Pioneer 10, it was so far away it was generating only a few tens of bits per second. This was a challenge only in terms of a weak signal, not in terms of the data rate. However, our thoughts about an extremely capable Mars Reconnaissance Orbiter in 2005 envisioned the projections of millions of bytes in a given day, perhaps even 1 billion bytes or more. This type of data rate would be much greater than any planetary mission in the past, and it meant that we had to have the right kinds of transmitters and we had to be prepared to receive this large amount of data.

Once we had rolled out the new program with all the missions up through the end of the decade, a new problem was identified by the DSN communications staff at JPL. A briefing on November 17, 2000, they'd had of the DSN pointed out that in December 2003 or January 2004, we were going in and out with "a traffic jam at Mars." What that really meant was that at that time the demands of existing missions, such as the Cassini mission on its way to Saturn, the landing of the two Mars rovers, in-place assets like the 2001 Mars *Odyssey* orbiter we were planning for, and the planned European Mars Express mission added up to a data rate that simply could not be supported by the existing capabilities. It is a little-known fact that the DSN serves the entire worldwide space community. There are antennas owned by other countries but only NASA and the United States had, at that time, a worldwide unified and integrated capability to track spacecraft anywhere in the solar system. It was very common in international arrangements for the Europeans to approach the United States and offer, for example, an instrument slot on a spacecraft in the future in return for the tracking and communications capability offered by the DSN.

Because of the way that NASA was organized in late 2000, the management of the DSN was not contained within the science organization at all. Rather, the communications capabilities that supported science missions and human exploration were all carried out in an organization that was devoted to space operations. This organization, the Space Operations Management Office (SOMO), was also the one that concerned itself with things like the former shuttle program—which was the biggest, most demanding fish in the NASA pond.

In this briefing in November, the DSN representative pointed out just how overstretched the system was. The load was sometimes 200 percent, twice the capability, or even at times 300 percent, three times the capability. While the requirements were still being revised, the picture that he showed made very clear that after the second rover was in place, the load on the network would increase even more. His "problem statement" was as follows: Based on current mission tracking requirements, the DSN does not have enough resources, that is, primarily antennas, to support the Mars view area mission needs in late calendar 2003 and early calendar 2004, even for the ideal case involving no antenna failures and no spacecraft problems. He went on to state that any deviation from the ideal case—that is, if any problem showed up—would be an aggravation and further increase the need for communications. His closing comment in stating the problem was that the situation could repeat every 26 months as missions approached Mars and conducted surface or orbit operations. From the Mars Program point of view, our desirable approach, which was to use every launch opportunity, was creating a severe systems problem in communications.

This traffic jam in the sky pointing toward Mars was a problem not just for the planned dual Mars rovers. It was also a problem for Cassini communications, Mars *Odyssey* support, the Japanese mission called Nozomi, the launch of a mission called Deep Impact, flyby of a mission called Stardust, and the anticipated flyby of a mission called CONTOUR (Comet Nucleus Tour). All of these were critical mission events, and about nine of them happened between November 2003 and February 2004. Clearly something had to be done.

The discussion of exactly what needed to be done, and more importantly, who was going to pay for it, continued throughout December 2000 and January 2001. This was the same time that I was in negotiations with Ed Weiler and with Steve Isakowitz at OMB over exactly what the Mars Program budget would be. In a briefing on December 21, 2000, just before Christmas, the people from SOMO laid out the results of their one-month study. They pointed out, once again, that when all the spacecraft are clustered in one area of the sky, the flexibility to equalize the loading is reduced. Luckily, I had some familiarity with the DSN through my work on Lunar Prospector and the work that had been going on at NASA Ames for decades with the Pioneer missions. It was generally regarded by the planetary mission community that the folks at the DSN were incredible colleagues and seemingly able to work miracles out of aging hardware. By continually upgrading the receivers and by becoming more and more sophisticated about scheduling both commands up and data down, they had been able to stretch the existing collection of 70-meter (230-foot) and 34-meter (112-foot) antennas much further than had been originally thought.

In the December 21 briefing, though, SOMO made it very clear that the science organization and the operations organization and the DSN people were all in agreement that the minimum mission requirements could not be met with current DSN antennas. The baseline solution proposed was that NASA would have to provide some basic upgrades and at least one additional 34-meter antenna at the site in Madrid to meet even the minimum requirements.

The bottom line at this meeting was that SOMO needed immediate authority to proceed with building this additional 34-meter antenna so as to meet the coming traffic jam problem in late 2003. They also said that they needed something on the order of $50 million over the next four years (the bulk of the funding was to be expended in fiscal years 2001–2003 with a small increment in fiscal year 2004). Everyone in the room turned and looked at me as the potential provider of this funding since a lot of the extra load came from the Mars Program. There were further briefings in mid-January, and this consideration became part of the budget that I was developing.

After negotiating with both Ed Weiler and OMB, it became clear that the Mars Program would have to bear this cost, because unfortunately the need for it was absolute. What I mean by that is that NASA and the science community were in full agreement about the value of the orbiters and the rovers, particularly in 2003, and it made no sense to spend hundreds of millions of dollars getting to orbit around Mars and to the surface of Mars and not be able to get the data back.

As we prepared the 2002 budget that would be announced by the president—whomever, according to the Supreme Court, that might be—in February 2001, the $50 million needed was included with my budget. We gave a tentative go-ahead to the space operations people and to DSN to begin discussions with the operating contractors based in Madrid. I remember visiting Spain and seeing the new 34-meter dish in 2001 as construction was begun. That new antenna became the workhorse to allow us to control the rovers to communicate with the orbiting spacecraft and to ensure that the major investment in the Mars Program was successful.

I would have to say that the media response to the rollout was positive, but we did get a lot of the questions that had been predicted, including why the Mars sample return, which was originally planned for 2003 or 2005, had now been pushed into the next decade. My answer to that, as it had been to the science community months ago, was that we wanted to understand the best possible place to get a scientific sample, and there were significant technological challenges that had to be met and developed before you could consider a successful sample-return mission.

Shown below is the actual press conference release that went out that day describing the series of missions. A decade or more after all the struggles, it is interesting to see that, while the European contributions mostly vanished, the U.S. program has stayed largely intact. There was a Scout competition that produced the *Phoenix* mission in 2007. The only other real change was moving the smart lander mission, now called the Mars Science Laboratory (MSL), which would carry a rover named *Curiosity*, from 2007 to 2009 to take advantage of a radioisotope power source. Subsequent technical difficulties, though, pushed MSL out to a 2011 launch with a 2012 landing.

NASA OUTLINES MARS EXPLORATION
PROGRAM FOR NEXT TWO DECADES
October 26, 2000

By means of orbiters, landers, rovers and sample return missions, NASA's revamped campaign to explore Mars, announced today, is poised to unravel the secrets of the red planet's past environments, the history of its rocks, the many roles of water and, possibly, evidence of past or present life.

Six major missions are planned in this decade as part of a scientific tapestry that will weave a tale of new understanding of Earth's sometimes enigmatic and surprising neighbor.

The missions are part of a long-term Mars exploration program which has been developed over the past six months. The new program incorporates the lessons learned from previous mission successes and failures, and builds on scientific discoveries from past missions. The NASA-led effort to define the program well into the next decade focused on the science goals, management strategies, technology development and resource availability in an effort to design and implement missions which would be successful and provide a balanced program of discoveries. International participation, especially from Italy and France, will add significantly to the plan. The next step will be an 18-month programmatic systems engineering study to refine the costs and technology needs.

In addition to the previously announced 2001 Mars Odyssey orbiter mission and the twin Mars Exploration Rovers in 2003, NASA plans to launch a powerful scientific orbiter in 2005. This mission, the Mars Reconnaissance Orbiter, will focus on analyzing the surface at new scales in an effort to follow the tantalizing hints of water from the Mars Global Surveyor images and to bridge the gap between surface observations and measurements from orbit. For example, the Reconnaissance Orbiter will measure thousands of Martian landscapes at a resolution of 20 to 30 centimeters (8 to 12 inches)—good enough to observe rocks the size of beach balls.

NASA proposes to develop and to launch a long-range, long-duration mobile science laboratory that will be a major leap in surface measurements and pave the way for a future sample return mission. NASA is studying options to launch this mobile science laboratory mission as early as 2007. This capability will also demonstrate the technology for accurate landing and hazard avoidance in order to reach what may be very promising but difficult-to-reach scientific sites.

NASA also proposes to create a new line of small "Scout" missions that would be selected from proposals from the science community, and might involve airborne vehicles or small landers, as an investigation platform. Exciting new vistas could be opened up by this approach either through the airborne scale of observation or by increasing the number of sites visited. The first Scout mission launch is planned for 2007.

In the second decade, NASA plans additional science orbiters, rovers and landers, and the first mission to return the most promising Martian samples to Earth. Current plans call for the first sample return mission to be launched in 2014 and a second in 2016. Options which would significantly increase the rate of mission launch and/or accelerate the schedule of exploration are under study, including launching the first sample return mission as early as 2011. Technology development for advanced capabilities such as miniaturized surface science instruments and deep drilling to several hundred feet will also be carried out in this period.

Mars missions can be launched every 26 months during advantageous alignments—called launch opportunities—of Earth and Mars, which facilitate the minimum amount of fuel needed to make the long trip.

The agency's Mars Exploration Program envisions significant international participation, particularly by France and Italy. In cooperation with NASA, the French and Italian Space Agencies plan to conduct collaborative scientific orbital and surface investigations and to make other major contributions to sample collection/return systems, telecommunications assets and launch services. Other nations also have expressed interest in participating in the program.

"We have developed a campaign to explore Mars unparalleled in the history of space exploration. It will change and adapt over time in response to what we find with each mission. It's meant to be a robust, flexible, long-term program that will give us the highest chances for success," said Scott Hubbard, Mars Program Director at NASA Headquarters, Washington, D.C. "We're moving from the early era of global mapping and limited surface exploration to a much more intensive approach. We will establish a sustained presence in orbit around Mars and on the surface with long-duration exploration of some of the most scientifically promising and intriguing places on the planet."

"The scientific strategy developed for the new program is that of first seeking the most compelling places from above, before moving to the surface to investigate Mars," said Dr. Jim Garvin, NASA Mars Exploration Program Scientist at Headquarters. "The new program offers opportunities for competitively selected instruments and investigations at every step, and endeavors to keep the public informed on each mission via higher bandwidth telecommunication on the web."

"NASA's new Mars Exploration Program may well prove to be a watershed in the history of Mars exploration," said Dr. Ed Weiler, NASA's Associate Administrator for Space Science. "With this new strategy, we're going to dig deep into the details of Mars' mineralogy, geology and climate history in a way we've never been able to do before. We also plan to 'follow the water' so that in the not-too-distant future we may finally know the answers to the most far-reaching questions about the red planet we humans have asked over the generations: Did life ever arise there, and does life exist there now?"

CHAPTER NINE

All Things to All People

Any time a significant new program is proposed through the budgetary process and there is an announcement about the details of the program, there is a well-understood and semi-formalized process in which you speak to all of the stakeholders before a public announcement. Everyone wants a piece of the action, of course, and wants to make certain that their particular wish list is fulfilled. An extraordinarily talented friend of mine at Ames, a senior NASA engineer by the name of Roger Arno, is also a wonderful artist and was in the habit of making deliciously satirical cartoons of various NASA folks and situations. He gave me a present of a sketch he had done, on his own time, of course, depicting the spot I was in. He has the contractors as vultures, which may be a bit harsh, but the general sentiment is not far off from how I felt.

Given all the attention that the administrator and, indeed, the world had put on the Mars failures, along with the new program being developed, our budget and mission queue certainly met the criteria for a major announcement. In the process leading up to the October 26 press conference where we announced the details of the missions beyond the two rovers in 2003, it was necessary not only to go and speak to OMB, but also to explain to Congress what we were planning.

There are two types of committees on Capitol Hill that take an intense interest in what NASA is doing. One committee, known as the House Committee on Science, Space, and Technology, has oversight responsibility, and every few years it will draft an authorization bill in which it lays out a broad perspective of what NASA shall be allowed to do and the amounts

Figure 12. Satirical look at the trials and tribulations of the NASA HQ program director with all the gates, traps, and obstacles to be faced. (Courtesy Roger Arno)

of money that should be allocated for NASA to carry out the charter. There is also a complementary science committee on the Senate side, the Committee on Commerce, Science, and Transportation. The real annual control of the budget, of course, lies with the appropriations committees. It is the appropriators who make the final decisions about the actual budget that any agency will have in a given fiscal year.

My job, sometimes alone, sometimes with Jim Garvin, was to go and speak to both the authorization and appropriations committees in both the House and the Senate. Unless there was an extremely important issue that was likely to come up for a vote relatively soon, the standard process was for an agency's senior staff person to meet with a congressional staff person. Often these congressional staffers are very young and very bright but not necessarily with deep background knowledge. Occasionally there is someone who's been in Washington a long time who really understands the details of the agency's budget and how it has evolved over time.

It seems that in Washington, especially in developing a new program, there is never time to do an extended set of briefings that occur over a period of weeks. It always seems as though everything is rushed and on a highly compressed schedule. Descriptions of $100 million or even $1 billion programs are compressed into hours and minutes. The Mars Program was no exception. While at this point, in late October 2000, we had spent perhaps seven months developing the new mission queue, the science rationale, and the budget, all of my briefings to Congress occurred in the span of a day or two around October 25.

My first stop was to a very friendly and thoughtful group, the House science committee, an authorization committee. The two senior members of this group, Dick Obermann and Ed Feddeman, had been around for years. Dick, in particular, was an exception to the young, bright, whippersnapper staffer rule. He had a PhD in physics and had been the Democratic lead staff member on the House science committee for a long time. Consequently, he was able to ask very good, deep, probing questions and also had a very thorough understanding of the complexity of space exploration. I described to Dick and Ed what the new program was all about, starting off with the fundamentals in our restructuring process, the accountability in the authority, what our scientific goals were, and finally the mission queue. By this time, I had become personally convinced that the likelihood of major participation from the French was small or even nonexistent. Even though they were still discussing collaboration, their primary interest was in Mars sample return. I had for several months believed that a Mars sample-return mission simply was the wrong thing to do at this time, and, given the French insistence that it had to occur before 2009, it was highly likely that their support would fade away. For the Italians I thought that there was a chance there would be some participation in the form of the communications orbiter and possibly an instrument or two. When I met with the congressional committees, the emphasis, luckily, as you might expect it to be, was on the U.S. portion. That was the element of the program Congress would be funding, and it was the one in which they were most interested.

My presentation package, after laying out all the constraints in the philosophy and so forth, focused on my "ladder to Mars," which began with an orbiter, Mars *Odyssey*, in 2001, the Mars Exploration Rover (MER) twins in 2003, the high-resolution Mars Reconnaissance Orbiter (MRO) in 2005, the competed opportunity Mars Scout Program, and then, in 2007, a long-life, long-range rover that would provide intensive *in situ* science and would set the stage for a sample return in the next decade. I didn't provide a detailed vision for the period 2012 to 2020. I left this next decade somewhat vague because the definition simply wasn't there to provide more than an outline of the possibilities. My approach was to talk about expanded access to the surface and subsurface, next-step decisions on human exploration, and sample-return missions. Part of the goal was to underscore my view that, given the necessary scientific reconnaissance and technological hurdles, a Mars sample return would occur only in the next decade.

The response from the authorizing committees was positive. They had questions about the support of the science community and the believability of our cost estimates but, in general, thought that a logical path had been laid out to the future. This logical progression with built-in resilience and robustness that emerged from a science basis pleased them a great deal.

The appropriations committee was another story. Appropriations committees, of course, had to deal with the annual budget cycle. That meant that they had to focus on what was being requested in the president's five-year budget and specifically what would be asked for in the next fiscal year budget, which had been prepared by the agency and approved by OMB. In October 2000 we were already technically in the fiscal 2001 budget, and the 2002 budget would shortly be passed back to the agencies and then negotiated prior to the president's State of the Union message. We were in a period of time known as the embargo. That meant I was not free to discuss any of the details of the 2002 budget and could only point to what the content was likely to be. This is always a very frustrating time when a senior agency official is meeting with a congressional staffer. There is a bit of a game going on. The staffer attempts to find out as much as he or she can about what's likely to be in the budget, and the agency person focuses on the content and tries to keep the discussion away from dollars and cents.

The appropriations group included the leadership of both the House committee and, most importantly, the Senate committee. On the House side, my chief inquisitor was Eric Sterner, who was educated in international space policy and was an up-and-coming bright person inside the Beltway. Eric and I had quite a lively debate. We covered not only the material in my presentation package but at times veered off into a rather arcane discussion of the orbits we were planning as well as the collaborations with the international partners. Eric, with his background in space policy and international studies, was keenly interested in what the Russians might do, for example. I told Eric that we had a trip coming up later in the year to go and meet with the Russians in person. I also noted that there was the possibility of both French and Italian participation, although my personal assessment was that parts of that were unlikely to occur. Sometimes, as with the other committees, there were questions that were submitted later on for us to answer. Sometimes these questions ran into the dozens and often would resemble the background "dirty questions" that we used in preparing for congressional testimony.

On the Senate side I got to meet with the well-known Senator Barbara Mikulski gatekeeper, Paul Carliner. Senator Mikulski was extremely well known as an advocate for NASA and the space program as well as for the Goddard Space Flight Center and for the Johns Hopkins University Applied Physics Laboratory (APL), both in Maryland, her home state. It was a truism that if you wanted to get Mikulski's support for any initiative, you had better be prepared to show Carliner what was in it for Goddard and/or APL. I must admit that because the bulk of the program would be carried out by JPL, I was at somewhat of a disadvantage in explaining how this program was going to benefit Maryland. Nevertheless, I pointed out that Goddard certainly had some good ideas on instrumentation, that they had provided many, many planetary space flight instruments, and that Lockheed Martin, with its headquarters in Bethesda, would undoubtedly be a participant as well. Carliner and his associates gave me a thorough wire-brushing, a good grilling on our plans for not only the future, but the next fiscal year. In the end, I think they were convinced that there was a logic, a method, and a scientific basis for what we were doing, and that if they supported the program, there was solid likelihood of a positive outcome. With the congressional briefings and those to the OMB behind us, our NASA Mars Program team made it to the official rollout on October 26.

Although the science community standing committees had reviewed the Mars Program several times in less mature forms before the rollout and would continue to look at it after our presentation, the first two reviews that occurred immediately after our public press conference were the most critical in terms of gauging the level of support that the new program would have.

NASA's science program has a set of standing committees that are advisory to NASA. These committees are set up under something called the Federal Advisory Committee Act and are required to hold all their discussions and deliberations in public and to advertise their meetings in advance. At the time that we presented the new Mars Program, the structure was as follows: The topmost committee was called the NASA Advisory Council (NAC). Underneath the NAC was a set of specialized advisory committees. The one we were most concerned with was the Space Science Advisory Committee. Without the support of the broader science community, we knew we would be dead in the water. Within the Space Science Advisory Committee structure there were subcommittees devoted to various aspects of space science. The subcommittee that concerns itself with planetary science was at the time called the Solar System Exploration Subcommittee, which was also crucial to have on our side.

The usual plan for these committees and subcommittees was to have, perhaps, three meetings a year, advertised well in advance, with a committee made up primarily of scientists, although the inclusion of an engineer or two, or perhaps a technologist, was beginning to become standard procedure.

The process was to brief the subcommittee and then the full Space Science Advisory Committee. It worked out that in late October and early November the subcommittee and the full Space Science Advisory Committee were all meeting together in California, and therefore the opportunity presented itself for me to make one more trip from the East Coast to the West Coast and be able, in the span of a few days, to brief both committees. While this was yet again more travel, it was a very efficient way to handle the briefings.

On October 30 I briefed the new program to the subcommittee, much as we had shown it at the press conference, and engaged its members with questions and answers for well over an hour. It was customary to go around the table and get a few comments on the major topics from each of the participants, and we did so. The first speaker was Ken Nielsen, an astrobiologist who was then at JPL. Ken noted that he thought the program was moving in the right direction. He himself was continually astounded by how little the community knew about what constituted life detection. He went on to say that he thought the sample-return mission should be very carefully scrutinized to be sure the sample was one that was very well selected. He didn't expect to learn much about life from a simple grab sample for just any rock taken from anywhere on the surface. He couldn't have done a better job for us if we had written his talking points.

Dave Swenson, a planetary scientist from Caltech, also endorsed the program, saying that Mars would continue to surprise us and that we should be careful about promising life detection because it's hard to prove that a particular instrument reading is definitively a sign of biology. Ellen Stofan, who was then at JPL, thought that the baseline program was very good. She was surprised at the cost increases over the promises that had been made in 1996 and 1998, but acknowledged that we were probably being more realistic this time.

Mike Drake, of the University of Arizona, who was the chair of the subcommittee, surprised everyone, I think, by characterizing the new program as "timid." I think that Drake expected that this reorganization of the program would somehow come out and endorse a very rapid sample return with very aggressive missions.

Jeff Cuzzi, a planetary scientist from Ames, liked the systematic way that the program would be developed, the technology inserted as it was ready, and he personally thought that the next-generation rover in 2007 would be worth a great deal. Dave Grinspoon, a planetary scientist from Southwest Research Institute in San Antonio, was very pleased with the program. He thought it would really "jazz the public."

Tom Krimigis, at APL, liked the science-driven, broad-based, traceable program—that is, a pedigree of science questions that led to the specific missions. Krimigis went on to say that he thought the *in situ* measurements of the landed missions would tell us a great deal and that that information could be used to make an intelligent selection for a future sample-return program. That was music to my ears and just what we hoped folks would realize.

Mike Zielinski, from Johnson Space Center, endorsed the idea of getting samples, but said that it was good to take your time to get the right sample. He noted that there would be a traffic jam with the Deep Space Network around the time that the MER twins would be landing. This topic was one that eventually required that the Mars Program pay a significant amount of the cost of creating a new dish at the Spanish station in Madrid. In the final accounting, the Mars budget paid for the entire $50 million that was required.

Laurie Leshin, most recently at NASA HQ but then at Arizona State University, commented that she thought the new program was thoughtful and measured, and she liked the science traceability and inclusion of funding for data analysis and instrumentation. She also noted that we had to do a thorough job of outreach to the public to explain this program and said she would like to see more preparation for the Mars sample return.

As each of these renowned individuals chimed in echoing our very own thoughts and conclusions, we felt that things were going really well. Finally, Bill McKinnon, a planetary scientist from Washington University in St. Louis, noted the following: Previous planning had been an orbiter and lander at each opportunity but with no clear purpose. He thought the new program was now rational and that the Mars sample return was now embedded in a logical framework. He thought that the program was exciting, not timid, and he saw no problem with public outreach. Phew!

As was customary, the chairman of the subcommittee, in this case Mike Drake, very quickly wrote a letter to NASA HQ summarizing the findings of their meeting. In it was one note, a short paragraph on Mars exploration. That paragraph said the Solar System Exploration Subcommittee received reports from Scott Hubbard, Jim Garvin, and Firouz Naderi. It was clear to the subcommittee that NASA had put its "A" team in charge of the Mars Exploration Program and overwhelmingly endorsed the mission set, laboriously constructed after consultation with a wide range of constituencies. Given budget constraints, the team had done an excellent job. A minority view felt the plan was "timid" in that it did not obviously engage the public and might be inconsistent with a possible future decision to send humans to Mars in the next decade. The timid comment, of course, came from Mike Drake, and he was the only one who expressed it, clearly exercising the prerogative of the chairman. I've had many discussions over the years with Mike about that comment, and he now acknowledges that he was too hasty in his assessment of the program. In some ways, though scientists are essentially skeptics by training, most planetary scientists love their work and really want to forge full steam ahead to get results for which they often have to wait a lifetime. Still, we had passed the first gate with more or less flying colors.

The next hurdle was the full Space Science Advisory Committee. They met at JPL on November 1–3, and we got the opportunity to brief them on November 1. The Space Science Advisory Committee consists of the chairs of the subcommittees along with a sprinkling of other scientists, and so this group included astronomers, solar physicists, and other researchers not connected with the planetary program.

In this case, the chair of the Space Science Advisory Committee was Steve Squyres who, as of October 12, had just been named the principal investigator of the payload on the *Spirit* and *Opportunity* MER twins. Steve quickly wrote a summary letter to Ed Weiler at NASA HQ and characterized the Mars exploration reformulation in the following way: "Scott Hubbard, Firouz Naderi, and Dan McCleese briefed us on the reformulation of the Mars Exploration Program. We were very impressed with the new Mars management team, as well as with the reformulated management structure which exhibits clear lines of responsibility for program planning and implementation."

The letter went on to say, "Scott described for us the anticipated structure of the Mars program for the next several launch opportunities. It is clear that code S. [the NASA science organization] has made significant progress toward establishing an achievable, science driven, technology enabled plan for Mars exploration. The plan is well-balanced, including the desired elements of orbital reconnaissance, *in situ* exploration, and sample return. The near-term program through 2005 is also characterized by a level of science return and technology demonstration that is consistent with the current level of funding. Because of the rapid pace of progress required to carry out this program, we recommend the science definition team (SDT) be formed immediately for the 2005 Mars Reconnaissance Orbiter and in the relatively near future, for the Mars 2007 opportunity." An SDT is a common way of being certain that the community is directly engaged in developing the scientific requirements and assessing how they are balanced against the engineering constraints and other concerns, such as cost and schedule. We could hardly have asked for more.

Steve Squyres's letter noted in bold and italics that an aspect of the plan that the committee endorsed with particular enthusiasm was the proposed Mars "Scout" line of missions. Each mission would be led by a principal investigator and have open competition. They would involve the broad community in the process of Mars exploration, would treat important problems with broader focus than the mainline program, and would enable the program to adapt a relatively short time frame to new discoveries. This was, of course, precisely what we had intended.

Steve and his committee continued, "We had a long discussion about the role and timing of sample return in the Mars program. Because of the challenging cost of sample return, we urge that Mars program managers articulate clearly the unique role that returned samples would play in addressing key scientific questions at Mars. We recognize the scientific importance of sample return, and we recommend the first Mars sample return occur as soon as possible, consistent with a prudent level of risk. We further recommend that the schedule for sample return be revisited, as appropriate, in the context of increased knowledge of the geological context of landing sites, potential future funding increases, and new technological developments. We endorse the approach of involving the scientific, technological, and human exploration communities in the future program planning." The letter went on to say, "It was also clear from what we heard that the Mars program offers an unparalleled opportunity to engage the public and students in the adventure of exploring another planet. It is important that the Office of Space Science take advantage of this opportunity to create a world-class education and public outreach effort as an integral and essential part of the program. We therefore encourage the Mars program leadership to involve leading figures from the EPO [education and public outreach] community, outside and inside of NASA, and to form a broad spectrum of alliances with the EPO community."

With the endorsement of the Solar System Exploration Subcommittee and the Space Science Advisory Committee, my team at HQ and the group at JPL felt that we had passed a major milestone and had secured strong support for our approach from the principal stakeholder, the science community. This set of reviews gave us a lot of confidence that we were on the right track, because even though there were many budget battles yet to come, Congress and the administration both pay serious attention to the opinions of the scientific community. The science community is generally regarded as being nonpartisan, factually oriented, and operating in the best interest of their constituencies.

We had told the public what we intended doing, and they seemed to approve. We had told the science community what we intended, and with more consensus than we had hoped, they too seemed to agree that we had done a good job and should move forward with the plans. Now all we needed was the money to do that.

In late 2000, the government was already well into fiscal year 2001 and preparing for the following budget, 2002. The final budget negotiations before the president's 2002 budget would be presented to Congress were carried out just as the holiday season was upon us, and it had been a very unusual year. The campaign between Al Gore and George Bush had continued well past voting day and ended up in the Supreme Court. This meant that the usual transition of power and the usual transition team activities were in limbo. Those of us in the agencies in Washington, D.C., were very unclear as to who would be calling the shots for the 2002 budget, which had been prepared under the Clinton administration yet would be presented under a new administration.

Very soon after our rollout on October 26 Brant Sponberg from OMB began contacting me in order to set up further discussions on the 2002 budget. Brant's first note was to inquire about the various reports in the media describing the rollout. He was a bit concerned that Mike Drake was critical of NASA as having a timid program and lacking vision, although the *Washington Post* article was very balanced. On November 1, Brant e-mailed to ask if there would be a possible time for us to go over the real details of the budget and have an intensive work session or two. I responded that yes, this would be possible; however, I had a commitment to go off to the international Mars working group around November 10 but did have some time at other points in November. I included that after December 7 I had promised my long-suffering spouse we would take some time off together. We planned to go into New York City via the high-speed Acela train and take the first vacation since I had become Mars program director.

Brant said that November 6 sounded good to him so he came over about two o'clock in the afternoon. While Ed Weiler encouraged us to work closely with OMB, he was nevertheless correctly suspicious about what the outcome of any given meeting might be. By this time Ed trusted me implicitly to have good judgment in what to tell the OMB people. Ed's main question was what would the product be of such an extensive discussion? I said it would be to support decision making for the 2002 budget in the next administration. We had the program vision already presented, but not all the budgetary details were pinned down.

The discussion with Brant went well. We identified various possibilities and various types of alternatives within the overall scheme that we had presented publicly. By this time in early November, we had settled on the *Odyssey* orbiter mission in 2001, the twin rovers in 2003, the Mars Reconnaissance Orbiter in 2005, and a commitment to some sort of very capable landed mission in 2007. Beyond that, things were less settled and depended in part on how we would approach the Mars sample return, and the role that the international partners would play.

Despite the importance of these OMB discussions, I felt obliged to continue to juggle aspects of the program itself, particularly the international piece. On November 7 I flew from Washington to Helsinki, Finland, via Frankfurt for a meeting of the International Mars Exploration Working Group that we in the United States called IMEWG ("immyweg") and the Europeans called I-MEWG ("I-myug"). The meeting was one that I deemed critical to attend in person, given the extensive discussions of international collaboration that were in play.

Our revised program plans were very well received by IMEWG. There were reports from each of the countries that participated. Marcelo Coradini, who then worked for ESA, proposed that they would be able to re-fly the Mars Express orbiter in 2005 although they weren't specific about exactly what they would do with it. This ended up being the Venus Express mission later on. The French space agency, CNES, surprised me and everybody else by talking about a sample-return mission to the Martian moon Phobos. I must say I did not appreciate what appeared to be an approach to negotiations by surprise tactic. The Russians were once again offering up a free launcher in support of the joint mission to Mars. I visited the Russian space agency November 29 through December 1 on yet another international trip, and we explored this offer. It turned out that the free launch included a number of integration costs that were sufficiently large that we would have been as well off buying the launch on the open market.

My old colleague Juan Perez-Mercader from Spain, whom I had met when I was setting up the Astrobiology Institute a few years before, offered that the Spanish space agency might be interested in providing an experiment or some type of communications capability. I noted the loading concerns on the Deep Space Network that were going on in the background. The United Kingdom requested support for the Beagle Program. This was a lander being developed by a single principal investigator, Colin Pillinger, without much assistance from anyone else, and he had run into funding problems. The Japanese were holding a workshop on their Nozomi mission to Mars, and there was a general discussion of the interest in human exploration by at least some of the countries. We set a date to hold the next IMEWG meeting at Kennedy Space Center just after launch of the Mars *Odyssey* mission in April 2001. My bottom-line promise to Ed Weiler and Earle Huckins was that I would prepare a list on both ASI and CNES that summarized all the quid pro quo for each country. Then I would do some sort of economic analysis on the deal and what their participation was really worth and if we had enough money on our side for the additional overhead.

Once I returned to Washington after this rapid trip to Helsinki, the top priority was to continue working with OMB on the 2002 budget request. The discussions going back and forth made it clear that maintaining multiple options was not really going to work for much longer. This attempt to be all things to all people was killing us. We had to settle on a point design and a single budget line and make decisions about what we were going to do in the future. By this point I was reasonably well convinced that OMB and the administration were very likely to agree to, and fund, our baseline program. We might also be able to get some additional technology money. However, it was becoming even clearer that the sample return was going to be off in the future and that we needed to prepare a budget line that was focused on the U.S. contributions, regardless of whether the French or the Italians showed up with large amounts of hardware.

Ed Weiler continued to work the problem by having off-the-record discussions with Steve Isakowitz at OMB. As promised, I took Susan into New York City for a few days around mid-December. We enjoyed the Radio City Christmas Spectacular, dined out at some interesting restaurants, smiled at all the fancy store displays and the rosy-cheeked shoppers, and generally enjoyed the Christmas celebration. It was a much-needed, much-appreciated break from being pulled in multiple directions by well-meaning but demanding groups.

Weiler had been pressuring me for several months to let him know whether or not I intended to stay in Washington, D.C., as the Mars program director for the indefinite future. Back in March when I met with Dan Goldin and accepted the assignment, the initial agreement was to come to fix the mess, full stop. We hadn't really talked about whether or not I would be staying in Washington for years to come. I promised Ed that I would think about this very hard while we were taking days off. It was during that time that Susan and I talked about the pluses and minuses of being in Washington, the immense stress of the job that I was doing, and the possible opportunities back in California. We had leased our house, not sold it, and most of Susan's family was in California. I loved California and my home center, NASA Ames. In the end, I think that this short vacation in New York was the time when, at some level, I decided that after returning to flight and getting the program funded and back on the rails, we would likely return to California. Although Susan was fascinated by the politics inside the Beltway and enjoyed watching all the news shows in real time, we were nevertheless isolated from everything except my job. When we returned to Washington about the middle of December, I told Ed in confidence, which he kept, that after everything was resolved—the budget established, *Odyssey* launched, and the new program under way—it was likely that we would return to California.

However, we were still not out of the woods by any means on the budgetary front. The people at NASA HQ and the science organization had been very diligently working with their counterparts at JPL and with me to figure out how much extra money we needed in order to carry out the program that we were proposing. As early as November 7 the HQ budget people, Roy Maizel, Voleak Roeum, and others, had shown me that if we took what we had then started calling the "balanced program"—which was the Mars *Odyssey* orbiter, the twin rovers, the support we had to provide for the European Mars Express mission, a robust technology budget that would be somewhere around 10 percent of the total, a Mars reconnaissance orbiter for 2005, and then a smart lander initially scheduled for 2007—we were still about $500 million short.

All this planning and budget preparation came to a head right before Christmas. I was contacted by Brant Sponberg asking if we had time to educate him and Steve Isakowitz and Doug Comstock on how these various options differed in terms of expected science and budget. I also communicated a series of crises du jour to Ed Weiler. The note I sent to Weiler on December 22 listed the missions that we were planning to work on through 2006. Let me note that, although the planning that was done back in August and September went out through 2012 or even later, the president's budget is a five-year budget. That means that what we had to address with OMB was the current year 2001 and the chunk of time and money that was 2002 through 2006. The budget that I sent to Ed Weiler on December 22 showed that if we stuck with a program that had the baseline U.S. effort up through 2006, using its operations costs and data analysis costs, we were about $555 million short for the next budget cycle.

It's worth noting at this point that in his direction to us, Ed was very clear that we needed to pay special attention to the data analysis piece. Space mission data are rarely straightforward. The information always needs to be correlated with spacecraft engineering data to ensure that you have it in the right context; it requires error checking and corroboration, and this is just to get started. The science results come from intensive analysis by teams of very smart, hardworking scientists who must bring all their expertise and experience to bear to ensure that we are seeing what we think we are seeing, sometimes in images but often in isolated data points and lines on spectra. Previous missions had not incorporated data analysis into their budgeting, putting the emphasis on the engineering and simple data gathering. It was always a scramble to come up with the money to fund the scientists to analyze the data we had spent so much effort in collecting. This time around we didn't want to repeat that mistake.

Over the brief Christmas break, all NASA HQ employees were told to go home. That is, they turned off the lights and turned off the heat in the HQ building from Christmas Eve until the day after New Year's. Of course, this didn't mean that anything actually stopped. It's just that people were then working from their homes and talking to each other on the phone. It was during this break that I got a call on my cell phone from Ed Weiler. Weiler pledged me to the deepest secrecy, but said that he had just had a conversation with Steve Isakowitz and wanted to know what we would be capable of doing if our budget got an increase over the next five-year period of something like $444 million. I assured Ed that we would have no trouble spending this money extremely effectively and that if he could get another hundred million dollars we would have the program in the bag.

Just before we all ran out the door on December 20 to take our Christmas break, Brant Sponberg contacted me wanting to know when we could come over to OMB and educate everyone on where the program stood and why we thought we needed additional funding. I responded to Brant on December 23 that January 4 to 5 would work the best. Brant and I both agreed that we ought to schedule an entire morning, or an entire afternoon, for this. Getting four hours of time from senior people in OMB is unusual, but it demonstrated exactly how important the Mars Program was to everyone.

The meeting that occurred on January 5 was the final pivotal meeting in answering the questions about what the new Mars Program was going to do, why it was strategic, and what the linkage was between the basic science questions and the missions that were planned. Ed Weiler did not attend this meeting. He had other pressing things to do, and by this point, he trusted Jim Garvin and me to represent the program. As we had done back in October, I laid out our strategies and plans in the most compelling way I could, including the "ladder to Mars" with the alternating orbital and landed missions, the lead-up to a sample return, and how to work with the Europeans whether or not various promises were fulfilled. I showed the latest budgetary discussions, and made the distinction between what would happen if we had to stay completely within the previous budget guideline and what we would obtain by extra funding.

Jim and I had worked out a tag team approach, with me describing the missions in the strategy, the schedule, and the budgets and Jim providing the scientific context. We carefully laid out Jim's strategy of narrowing our possible landing sites methodically, based on life questions as the primary driver, from the thousands of spots on Mars that might be of interest based on Mars Global Surveyor data to the small focused number we would need to justify a flagship-type mission or, ultimately, the holy grail of sample return. We showed how the *Odyssey* mission in its orbit would look for evidence of water ice and understand the mineral distribution better than Mars Global Surveyor. We pointed out how the twin rovers in 2003 at two different scientific hotspots would provide us ground truth for the MGS and *Odyssey* observations. We pointed to the Mars Reconnaissance Orbiter and how the instruments that we expected to be proposed would give us a much-higher-resolution picture of the surface of Mars and help narrow the places of interest even further. We showed the power already demonstrated by the Discovery Program of having a competed mission in 2007. The underlying message, of course, was that all of the pieces fit together to make a wonderful whole. We needed a full, uncompromised budget to make this all a reality.

We kept emphasizing that although everyone seemed to want sample return eventually, the emerging consensus from the science community was that we really needed to understand where best to go to get a sample that was indeed worth $2 billion and at least six years of effort. Jim provided a very compelling graphic for all of this, showing the different phases of Mars exploration, beginning with the thousands of interesting spots from the MGS observations, then down to hundreds, down to dozens, and, by the end of the decade, with the missions we had proposed, down to a handful of scientific hotspots that would be the most compelling sites for a sample-return mission.

It was at this point that Brant asked me a question that you would only hear inside the Beltway. He pointed out that we had just had an election and they wanted to know if they approved the 2007 long-range science laboratory for the surface of Mars "could it discover life before the end of the second term of the president?" Mentally my jaw dropped. I had been coming into contact with the political driving forces of Washington, D.C., for decades, but I'd never heard it put so baldly. Inside my head I was laughing and groaning at the same time, but I responded with, "Well, here's a possibility. Let's say that our 2007 lander has on board an electron microscope"—one of the possibilities we'd been discussing—"and what if the 2007 lander obtains the sample, looks at it with the electron microscope, and sees evidence of the same kind of Mars micro-fossils that the Allan Hills meteorite showed only it will be *in situ* at Mars. What about that?" I asked. This seemed to satisfy them that we had thought through our "follow the water/looking for life" theme to a proper political conclusion.

By the time Jim Garvin finished his usual outstanding job of describing the science strategy through 2008, I believe that Steve was convinced that we had a series of missions that would each provide rich scientific return and significant public engagement. We took pains to explain to Steve that we had not created any new information for this presentation but that all this was thought out to reach the mission queue and technology plan in the first place.

In reviewing the options, I first reminded Steve of the schedule issues that we needed to address in both properly inserting technology and feeding back discoveries into subsequent missions. I thought this would be old news to Steve, but he seemed genuinely surprised by the time it takes to go from an investigation idea through technology and mission development and then data return to pose new research questions. The mission pacing we adopted in our new plan became more clear and made clearer why the old plan was undoable.

At the conclusion of our meeting I went back through the budgets and what we had and what we didn't have. I showed that while we were proposing something beyond the old budget from previous planning, we were offering a more robust Mars Program, plus the additional funding would allow us to upgrade the DSN. I had explained to OMB that without this additional antenna we would simply not be able to send up the commands to handle all the spacecraft landing and orbiting Mars and get the data back as planned. Also, with our request we would be able to provide full funding for a capable Mars Reconnaissance Orbiter. If we had to stick with the previous budget guidelines, we would not have sufficient funding for the instrumentation that we thought was necessary to go to the next step of orbital reconnaissance.

Next, we were still planning for a telecommunications orbiter provided by the Italians, and I had concerns about whether or not we had adequate interface money on our side, assuming the Italians would follow through with their commitment. It was important that we be able to ensure that the systems engineering in the whole program integration was adequate. Assuming that the French would provide an orbiter with their own experiments on board, they still needed a type of heater called a radioisotope heater unit. The French did not make these. Only the United States was capable of providing them, so some amount of money had to be set aside for that. Next, and very importantly, our plan included a smart lander for 2007, which eventually became the Mars Science Laboratory. For the years at the very end of the decade in 2009 we saw the possibility of another type of orbiter using a radar system, and we were planning for the beginnings of a sample-return mission in 2011.

I think we were reasonably successful in answering all the questions and giving them everything that they wanted to go off and develop a consistent budget strategy. It was important that we keep reminding ourselves that Steve Isakowitz and his team had responsibility not only for NASA's budget, but also for the science budget of the Department of Energy and the National Science Foundation. We were competing against other scientific priorities, but I think that I detected in Steve and Brant and Doug a real interest in the Mars Program and what we could do to advance our understanding of not only the physics of the solar system, but the very big questions of whether we are alone and whether life ever emerged anywhere else.

As soon as I returned to NASA HQ on Friday, January 5, after our meeting with Steve and his team, I wrote Ed Weiler a short synopsis of our visit. I first noted that I believed we had met our objectives for the meeting, which were to give a detailed and exciting science story for the missions in the budgeted program and to provide the key discriminators that showed what the various out-year budget options would provide. As we had guessed, Steve had to explain to his bosses why this new program was as exciting as the old, undoable program with its landers and orbiters at every opportunity and its early sample return. Furthermore, why should it get even more money? Steve reminded us that under the old program, the simple explanation of samples returned to Earth by 2008 to look for life was sufficient to argue for funding. In the new program, sample return was beyond the end of Bush's second term, so what were they getting? Our job was clearly to take the emphasis off near-term sample return and point out the incredible science return of the next few missions, and how they would take a major step toward answering the big questions.

I told Weiler that the comprehensive budget as we presented it met Steve's expectations for a major Mars campaign, should anyone ask. Our bottom line and actions after this meeting were that Steve said we had met his requirements and seemed genuinely pleased with the much clearer picture of the anticipated return from the new program. This told me volumes about how important the rollout of any new program is to the stakeholders, decision makers, and gatekeepers. Steve was particularly pleased with the science strategy and said that Jim Garvin had cleared up the piece he had been poking at Brant about. Our action was to prepare a white paper to expand on the mission content and to complement the view graphs we had used to present our plans. It's a fact of life in Washington that no matter how many times you explain things and no matter how eloquently, you are regularly asked to explain the whole thing again in a slightly different way.

We also were asked to identify the technology plan and point out where we were making assumptions about co-funding or leveraging other programs. Steve was especially critical of the previous program, which apparently never delivered a well-thought-out technology plan. Our action was to provide a technology plan in the near future. I had anticipated this and required JPL to come in the next Friday, January 12, and present us with the latest plans in at least "near-final-draft" form. The plan included 50 percent spending outside JPL with cooperation from NASA Langley, Marshall Space Flight Center, NASA Ames, and Johnson Space Center, something guaranteed to make us popular with NASA's ninth floor.

We were also asked to identify areas for the Mars Program that were vulnerable and could need funding out of the requested augmentation. I identified four areas immediately: *in situ* instruments, should the instrument program not go ahead; radioisotope power for Mars, should another program take one of the few radioisotope units left; extra funding for the 2005 Mars Reconnaissance Orbiter to ensure advanced instrumentation for the 30-centimeter resolution level; and funding for the Italian instrument. While we still didn't have a firm commitment from the Italians, if their telecommunications orbiter came through, we would reciprocate with an instrument.

My final note to Ed about our meeting was that although highly complimentary about the presentation, Steve was very cautious about the future. He said he had no idea what the new administration would do and suggested that we might be lucky to hold on to our present budget much less get an augmentation. He advised us to plan for the "in-guideline" budget, the old one, for now, maybe with some small additions but to stay tuned for developments.

The period from early January through early February was filled with a lot more activity on project definition, budget exercises, and so forth. In particular I got a note from Brant Sponberg on Wednesday, January 17, asking us to get him a ten- to twenty-page white paper, a set of charts, as well as some budget items by Friday, January 19. Things often move very, very quickly in Washington, but you also need time to think through these very complicated and detailed science and engineering issues. Fortunately, I had anticipated a great deal of this, and with the help of Jim Garvin, the staff at JPL, and my own team, we were able to deliver everything that would be wanted.

Finally I got a note from Ed Weiler on February 9 labeled "your eyes only" saying, "Here is the fiscal year 2002 budget run-out with current OMB guesses," Ed's personal term. That budget had the full amount to cover the DSN antenna cost, the additional funding for the missions, the support for the smart lander in 2007, and more. This showed a total increment of $535 million over five years. We could do it all and do it well! I didn't break open the champagne just then because Ed had warned me that until the president officially unveils his budget, things can change or disappear. It's best not to count your chickens until the check clears the bank.

On February 28, the president's budget was unveiled. The celebration in the Mars Program office could be heard to the end of the hallway at 300 E Street SW. We had received, over the period from 2002 to 2006, a total increase of $548 million, or more than $100 million a year total. The words in the president's budget sounded like they came straight from our view graphs. "The newly reformulated Mars Exploration Program is pursuing four major goals and objectives: (1) Determine if life ever arose on Mars and if it still exists today, (2) Characterize Mars ancient and present climate and climate processes, (3) Determine the geological processes affecting the Martian interior, (4) Prepare for human exploration of Mars, primarily through environmental characterization." Now it was time for the champagne.

The budget words continued: "The newly restructured Mars Exploration Program will deliver a continuously refined view of Mars with the excitement of discovery at every step. The MEP strategy will respond to new science investigations that will emerge as discoveries are made. Strategy is linked to our own exploration experience here on Earth and uses Mars as a natural laboratory for understanding life and climate on Earth-like planets other than our own." Playing back our scientific approach, it said, "The basic scientific approach to achieving these goals is one of Seek, *In Situ*, and Sample." The president's budget made note that we would be doing orbital reconnaissance and then confirming from the ground the observations. Finally, it said that samples of scientific significance would be returned to Earth for investigation in ways that are not possible to be performed on the surface of Mars.

The funding profile matched nearly everything that we had requested. The document went on to outline the specific missions, the ongoing operations of Mars Global Surveyor and Mars *Odyssey* in 2001, and the 2003 Mars Exploration Rovers. There was a budget line for future Mars exploration that included European collaboration, the 2005 Mars Reconnaissance Orbiter, and the 2007 smart lander. In addition it noted that a 34-meter new dish for the Deep Space Network would provide the communications requirements as needed. To say that we were happy with this budget is an understatement. After almost a year's worth of very hard work and formulation, building support, narrowing down possibilities, and working with the industrial, international, and political communities, it yielded the resources to go and do what we said we were going to accomplish.

No rest for the weary, my mother used to say, and she must have been right. Shoehorned into this same period was our trip to Russia, in the middle of winter.

The engagement of the Russian Federal Space Agency, commonly called Roskosmos, began in the summer of 2000. By that time we had the outlines of the entire new program beginning to be well understood, and we had had some considerable interaction with our French and Italian and ESA partners. However, I was encouraged by the people in NASA's international affairs office to at least consider whether or not Roskosmos could be involved.

The Russians, of course, had tried to orbit Mars and land on Mars many times but had never been fully successful. The very earliest attempted mission to Mars was by the Russians around 1960. This mission failed. The Russians tried over a dozen times to explore Mars with robotic spacecraft, and according to the records assembled by some European historians, the best that ever happened was the return of a few images from the surface of Mars and a few images from orbit. Any interaction with the Russians had to be limited to an area where they had proven expertise, such as launch vehicles.

The Russians had clearly heard about our replanning efforts and contacted Dan Goldin and raised the issue of whether there were cooperation opportunities. So on July 13, 2000, I wrote a letter to Georgi Polischuk, roughly Ed Weiler's equivalent in Roskosmos. I briefly described our planning and solicited their possible engagement. A series of interchanges took place at various levels where the Russians expressed interest in working with us and suggested that we get together and talk.

As often happens, discussions between countries drag on and are delayed by a variety of factors. The next substantive interchange occurred in October 2000 in which Ed Weiler informed his counterpart that there would be a visit in late November at which we would discuss a variety of technical cooperation possibilities including a Mars subgroup. The culmination of all of this interaction was a trip leaving the United States on November 27, arriving in Russia on November 28. The meeting occurred between the Russian counterparts and us on November 29 and 30 and December 1. This was an unforgettable meeting in my memory.

It was my first exposure directly to the Russian culture, although I had taken Russian in undergraduate school. The preliminary agenda called for me to give an overview of our Mars Program after lunch on the afternoon of the first day. I had heard enough about predictable Russian lunch activities to request that I be allowed to give my talk in the morning before lunch. This proved to be a wise decision.

We arrived in Russia in winter, and everyone was bundled up in fur coats. There was snow on the ground and smog in the air, and we were transported to the hotel that was being used to house the constant stream of engineers and managers developing the International Space Station with the Russians. An entire floor of this hotel, which had gone by several names—I believe it was the Renaissance Hotel when I was there—was devoted to NASA personnel. In fact, we even had our own NASA telephones at bedside next to the local Russian telephones. I recall picking up the NASA telephone and hearing a delightful southern accent say to me, "This is Marshall Space Flight Center. What can I do for you?"

We left the next morning in our bus with our staff and our foreign international relations people for the Babakin Institute. The Babakin Institute is the Russians' front door for science investigations to the rest of the world. Later on we would visit the Lavochkin plant, which is more nearly like our Lockheed Martin. Once we had exchanged introductions and pleasantries around the table, I gave my overview of the new NASA Mars Program. This was done through sequential, not simultaneous, translation and therefore was very slow to deliver, but with the help of my "haiku" approach—simple and clear with as few words as possible—I got through it and, I believe, communicated reasonably effectively our plans for the future. The Russians asked me questions, and we had some discussion then and there about what possibilities there might be for collaboration.

Next we went to our first Russian lunch. It was as I had imagined from movies and stories from international travelers. On the table were plates, knives, and the usual tableware plus a bottle of vodka for every two people. We began with a toast led by our host, the affable Constantine Pinchkadze, director of the Babakin Institute. Needless to say, one must respond to a toast with a toast. By the end of the lunch we had consumed a substantial amount of the two-person vodka bottles and were feeling quite happy for two o'clock in the afternoon. At this point any reasonable person would want to take a nap. However, the Russians served us Turkish coffee, which had the consistency of 30-weight oil and was the color of dark milk chocolate, and side orders of sugary treats. Powered on strong coffee and cookies, we made it through the rest of the afternoon in our splinter session discussions.

There were tours of both the Babakin Institute and the manufacturing plants. We got to visit their Museum of Space hardware, and I was impressed by the bulletproof nature of their designs. By that I mean that every structure seemed to consist of steel and inch-thick bolts and rivets like those you might associate with the Golden Gate Bridge. The tour of their space museum was led by a man whose claim to fame was that he'd bolted Yuri Gagarin into his space capsule on the first-ever manned trip into space.

Leaving the Babakin Institute late one evening, Jim Garvin noticed young women in fur coats standing alongside the road, seemingly with bare legs, and there was some question about what they might have on beneath the fur coats. Jim asked me why these girls were standing there. Weren't they cold? I explained to Jim that these were most likely working girls who might provide some service to the engineers of the institute who were on their way home. Sometimes even the most brilliant scientists seem to need a bit of help with the more pragmatic sides of life.

The lunch, however, was not the high point of our social visit to Russia. That peak was reserved for the dinner that we had the evening of the second day. As we made our way in the van to the restaurant I noted that there were large numbers of Mercedeses and other expensive cars parked in the slush and dirty snow outside. I also noticed that there was a guard carrying a Kalashnikov submachine gun marching up and down outside of the restaurant. I inquired of both our Washington representative, Diane

Rausch, and our guide the purpose of this gentleman. It was explained to me that restaurants in Moscow were sometimes the object of relatively hostile takeovers. In this particular case, the restaurant had been owned by Russian mafia group A. Russian mafia group B decided they wanted to own a restaurant, so Russian mafia group B came to the restaurant, killed the owners, and took over. The guard, an employee of Mafia group B, was posted there just in case any survivors of group A decided to revisit the restaurant. I was starting to look forward to the coming vodka.

Once inside the restaurant it became clear that we were in a theme establishment. The name of the restaurant was taken from one of the most popular Russian movies, *White Russian Nights*. In this movie the brave Russian revolutionary arrives in North Africa or the Middle East and discovers women trapped in a harem. The brave revolutionary frees the harem girls from their oppression and establishes a new republic of the proletariat, or at least this was the story that I was told by our guide. It was clear that the owners of the restaurant took this movie theme quite seriously. The drapes inside the restaurant all were created to resemble a harem of the Middle East. The waiters were dressed in costumes evoking Arab and North African clothing, and the tables were set to be consistent with the restaurant's overall Moroccan or Middle Eastern design. There were perhaps fifteen of us. We were divided into two tables. I was seated at the table with the leaders Constantine Pinchkadze and Georgi Polischuk. Also at my table was my good friend Gentry Lee.

We began the evening with some good Russian beer and an assortment of delicacies including sliced meats and black bread. Then, of course, it was time to toast everyone with vodka as we had done at lunch. Along with the toasts we were served some really good Russian caviar. The vodka was properly chilled, ice cold and viscous. It froze your tongue on its way down your throat. Another course of delicacies was served. Then they brought out the Georgian red wine. The wine was in large earthenware jugs and was really quite good. There was a fish dish, and then we had a small break.

During this break we learned that every half hour or so belly dancers consistent with the theme of the restaurant would come to entertain us all. There were perhaps a half dozen of these scantily clad women, who were clearly very expert in the art of belly dancing. One of these noted Gentry's interest and came over so that her gyrations were mere inches from his face. It took some time before everyone's attention returned to the meal. If they were trying to soften us up for something, they were doing a perfectly splendid job.

Next we had a meat dish, more Georgian red wine, another dancing break, and then finally we moved to the dessert trays. More of the Turkish coffee was served, and perhaps three hours or more after we had entered the restaurant, dinner was over. I did note that although Constantine was eagerly entering into every drinking and eating course, Georgi was much more reserved. Over dinner Georgi probed to see what my level of interest might be in having Russia as a partner. I was careful not to commit to anything, especially given the heavy vodka consumption we all, with the exception of Georgi, partook of during our meal.

The next morning it took some doing to have everyone ready to go at 7 a.m. Gentry, uncharacteristically, was late and appeared to be just slightly fuzzy around the edges. Years later Gentry told me that in all his professional life that was the only time he had ever been late for a meeting. Clearly the evening's festivities had taken a toll.

It was only on the final day of discussions, after all the presentations and tours, that I was brought to meet Yuri Koptev, the head of the entire Russian space agency. Koptev personified the Russian bear—a very tall and heavyset man who loomed over you. Luckily, I had been somewhat prepared for this meeting through advance work done by our international affairs staff. With a flourish Koptev produced a document, mostly in Russian, that promised three launches to NASA as part of a partnership that would ultimately result in a sample-return mission. I was assured that all that was required would be some minimal integration costs. I received the letter with appropriate diplomacy and promised a thorough assessment.

It was in the drive leaving the Babakin Institute and heading back to our hotel that one of the Russian staff spoke to me about the details of this offer. In this discussion I learned that the integration costs probably would amount to the price we might pay for a launch vehicle if we simply bought it on the open market from the Europeans. So this was really not a freebie at all but rather a way for the Russians to get hard currency, which they desperately needed at the time.

In the end, we pressed ahead with the new Mars Exploration Program with no Russian participation. The sense around us at HQ at the time was that there was no money to pay the Russians for launches, and even if we had had the money, it would be a very politically difficult thing to do. So ultimately, we didn't see any other mechanism whereby they could provide something that would be of great value in the near term. Even now, more than ten years later, the Russians are still trying to reenter the Mars game with their Mars mission called Mars "Grunt," a Phobos sample return. *Grunt* is "soil" in Russian.

In some regard, creating the new Mars Program had been like opening a set of Russian matryoshka dolls, those elaborate sets where every time you open one and think you've come to the end, there's another one hiding inside. Still, we felt as if we had our dolls all opened and lined up in a row now. We had the right people. We had our organizing principles. We had our communications tools. We even had our missions, and now it looked as if we had the money. Of course, it wasn't long before there were numerous additional adjustments to the budget, but for a brief moment the whole team at NASA HQ, JPL, and the rest of the community could bask in the glow of a restructuring job well done. Now, of course, we had to deliver. The Mars *Odyssey* mission had to be successfully launched, the Mars rovers had to be successfully developed, the Mars Reconnaissance Orbiter had to be defined and its instruments acquired, and a 2007 lander had to be given shape and definition.

CHAPTER TEN

Return to Flight

A Mars Odyssey

Planning is all well and good, but in the spaceflight world, it's all about firing rockets. No matter how careful we are, no matter how huge the armies of people or how dedicated their hearts and minds are to the process, actually being ready for launch and firing the rocket that starts you on your way is everything. That ignition is the beginning of the real-world experience that will either justify or negate all your efforts. Or course, it's really only the first step in a grueling, nail-biting series where failure can sneak up on you from myriad directions. Still, though, that first moment feels and is dramatic. There's smoke and fire in a launch, and the drama is palpable. For the Mars Exploration Program, our "return to flight," arguably the true beginning of our program, was the launch of Mars *Odyssey* from Cape Canaveral. A decade of hopes and dreams was at stake. We all wanted this to succeed, and I found myself in the driver's seat.

The *Odyssey* orbiter mission to Mars that launched on April 7, 2001, had had a very checkered history. The mission was first created during the faster, better, cheaper era of the late 1990s. The spacecraft and the designs that were used, as well as the engineering and management approaches, were all part of the same group of missions that included the failed Mars Climate Orbiter (MCO) and Mars Polar Lander (MPL). As a consequence, the Mars 2001 mission got intense scrutiny, both early in 2000 and then continuing right up to the day of launch.

In addition to all of the customary reviews that a flight project goes through, *Odyssey* was subjected to literally dozens of additional examinations, looking at nearly every part, nearly every design feature, nearly every drawing, to try to figure out whether there were any fatal flaws in the mission such as the ones that caused the loss of the two previous Mars missions, MCO and MPL. Out of all these reviews there were, in fact, a few very serious findings and recommendations.

The first problem was one that dated back to the ill-fated Mars Observer mission that had disappeared in 1993. There was a serious concern that a type of common spacecraft propellant called hydrazine could explode while in a fuel line during the 80 seconds or so of time when it is being pressurized, immediately prior to the ignition that would allow the spacecraft to go into Mars orbit. This would cause a line rupture and loss of the entire mission. This scenario is the most likely cause of the loss of Mars Observer, although no one can ever say for sure. It was the "most probable" one identified by the failure review board. The recommendation to the Mars *Odyssey* project was to install a device called a check valve, a piece of hardware that prevents any type of fluid from flowing backwards. It's a very common device even in household water lines. The initial response in July 2000 of the project office, not surprisingly, was to evaluate whether installing a check valve so late in the mission would create more problems than it solved. When these different recommendations were being considered, the project was already well into its assembly and test phase, and it is a given in this tightly constrained, risky space world that once the design is frozen, you had better know what you are doing if you change it, even a tiny bit. A complete spacecraft had been designed and built and was going through the very late stages of processing leading up to the shipment to Cape Canaveral and subsequent launch.

There were a few other serious concerns identified in the first half of 2000. The spacecraft had two fairly standard-sized fuel tanks as part of its design. Fuel equals life to a spacecraft. One concern was that the center of mass of the whole spacecraft might shift due to the fuel transfer between these two tanks, which would result in wobbling and an inability to properly control the positioning of the spacecraft as it was going into Mars orbit. Fixing this would also require additional valves and the project had to consider, once again, if risk was equal to the reward. Parenthetically, the review boards noted that all six of the valves that would allow for the orbit insertion had to open successfully or the mission would be lost. There was no backup valve as there had been for other large missions like Viking, Galileo, and Cassini.

There was also much concern that the solid-state power amplifiers would not survive the planned usage. Some 9,000 on/off cycles were planned throughout the mission. The recommendation was to change the operating plan to reduce the number of cycles—that is, leave it on unless there was an absolute reason to turn it off.

Finally, there was a whole series of issues about whether or not there were enough of the right people working on the project, particularly in the software area, but also in the areas of planning for various contingencies and addressing something called "fault protection." Fault protection is the term that's used in the industry for creating a response to some anomaly on board the spacecraft. If you ever hear an announcer say that a spacecraft has gone into "safe mode" or if you read this in the media, what it means is that the fault protection program has been activated on board because some unusual or unplanned signal has been noted. That gives the operators on the ground the chance to read out the data, figure out what's happened, and send a new command to fix it—knowing that the spacecraft is in "hunker-down mode" and is not doing anything that might create further problems.

There was another major review of the project in September 2000. And then there was the pre-ship review between December 4 and 6, the standard review where all of the spacecraft, instrumentation, and test results are reviewed to see if everybody believes that the hardware and software are ready for shipment to and integration with the launch vehicle at Cape Canaveral. Although there were still concerns that had not been addressed, the results of the pre-ship review were positive, so I concurred with George Pace, the outstanding project manager for Mars *Odyssey*, that it was time to go to the launch site.

Many people associate launches in Florida with the Kennedy Space Center (KSC). Certainly that is the place where all the human flights take place. The shuttle pads and all the associated hardware are under the control of KSC. However, the robotic space science missions, commercial missions, and military missions all launch from the Cape Canaveral Air Force Station. While I was grappling with the administration, the agency, Capitol Hill, the science community, and so forth to create the vision and find the funding for the Mars Program after 2001, there was all of this project work going on at the Jet Propulsion Laboratory, Lockheed Martin in Denver, and at the Cape. Such was the job I had assumed.

Details matter in all endeavors but rarely more so than in space. Careers are on the line with missions, and folks grow to care about them as if they were family members. With the level of emotional investment brought on by long hours and intense teamwork, even the name matters. In August 2000, we had addressed the issue of what to call the first new MEP mission at NASA HQ. Up to this point, the mission had simply been known as the

Figure 13. Odyssey project manager George Pace in the VIP visitors gallery that looks directly into JPL Mission Control, explaining the process of firing the engines to (successfully) insert the spacecraft into orbit around Mars. The board to Pace's left shows the status of the Mars orbit insertion.

Mars 2001 orbiter mission. A series of candidate names for the 2001 Mars Surveyor mission were solicited via the JPL project manager from the JPL project staff, the project science group, and the project's prime contractor, which was Lockheed Martin. Over 200 names were submitted. The final candidate names were coordinated by a naming committee consisting of Mark Dahl, the HQ Mars 2001 program executive; Michael Meyer, the Mars 2001 program scientist; Steve Saunders, the project scientist; and Don Savage, the representative of public affairs assigned to the 2001 mission. Believe it or not, NASA has a policy directive that instructs the HQ staff on how to go about selecting a name for a mission. The result of all this effort was to call the mission the Astrobiological Reconnaissance and Elemental Surveyor, or ARES. Ares, of course, was the Greek god of war, the Roman version of which is Mars. Ed Weiler recommended to the associate administrator for public affairs that, as of August 31, the name be ARES.

This was the duly announced. Not everyone was happy. There were members of the project and members of the public who thought that ARES might not be the best choice. It was not very compelling, was awfully aggressive, and as an acronym was pretty nerdy. A number of people, myself included, had noted that "2001" was contained in the famous name of Arthur C. Clarke's seminal science fiction book *2001: A Space Odyssey*. The naming committee got back together and reviewed everything one more time and noted that one of the candidate names, 2001 Mars *Odyssey*, had been considered but rejected because of trademark and copyright concerns. In one of the fluke easy things that does occasionally surprise us in this usually excruciating business, by contacting Sir Arthur C. Clarke in Sri Lanka directly, with a simple e-mail, NASA learned that not only did Sir Arthur have no objections to naming the mission after his famous book, he was, in fact, delighted with the idea. On September 20, Ed Weiler wrote another letter to the associate administrator for public affairs, who was now Peggy Wilhide, recommending a name change from ARES to 2001 Mars *Odyssey*. As Ed's letter notes, this change had "greater potential to engage the public and inspire our education and outreach efforts." The letter went on, "If you concur in the name change, the Project Office and Project Science Group will be notified, and the name will be used in future press releases and public affairs efforts." On September 21, Peggy signed the note and wrote in the margin of the letter "Love it!" with a big circle around it.

Another little-known feature of NASA's mission process is the requirement well before launch to create a document called the mission success criteria. This requirement ensures that no one can either accuse NASA of changing the rules after the fact or claim success when there was none. In March 2001, I received the document that had been prepared by George Pace at JPL, marked up at HQ, signed by the program scientists at both HQ and JPL, and finally signed by Ed Stone, the director of JPL. What it said was that the Mars *Odyssey* spacecraft would be launched in April 2001 on a mission to orbit Mars and carry out a global remote sensing survey for one Mars year, which is roughly two Earth years. It goes on to describe when the mission would arrive at Mars, what kind of orbit it would be in, and the payloads. The spacecraft was also required to provide for the relay of data from the surface elements landed by the United States and other nations for an additional Mars year.

The science objectives are spelled out at a level sufficient that almost anyone could determine whether or not the mission had been a success: for example, globally map the elemental composition of the surface; acquire high spatial and spectral resolution mapping of the surface minerals; determine the abundance of hydrogen in the shallow subsurface. The ultimate objective, of course, was extremely important. There were such strong indicators that there had once been water on the surface of Mars that we wanted to figure out where the water went. Was there, in fact, as some people speculated, a lot of water tied up in ice or would we find that, except for the caps, Mars was a very dry planet? The science objectives went on to say that there should be information on the makeup and texture of the surface. All of these objectives had one eye toward eventual human exploration of Mars, including characterizing the Mars near space radiation environment.

This document is very short, only four pages. Two of the four pages are signature sheets, so there are only two real pages of content. After the summary of the mission and the science objectives, the document goes on to describe complete success criteria and minimum success criteria. The first element of success, of course, is to get the spacecraft in orbit around Mars. Complete success requires that it be in the optimum science orbit, just for starters. The second item of complete success was to carry out a global survey of Mars for one Martian year, 687 Earth days, and collect 75 percent of the science data that can be nominally returned by the instrument complement. Included is the requirement to analyze these data sets in the context of the science objectives above and disseminate the results to the community, plus provide relay communications capability for surface elements up to one Martian year following the survey mission, and to archive the data in the planetary data system, after a short—less than six months—period for calibration and validation. Finally, we had to provide for regular public release of imagery and other science data via the Internet, as well as regular releases for public information purposes. NASA is very serious about metrics.

Minimum success was to insert the spacecraft into orbit around Mars. In other words, even if we didn't reach the optimum science orbit, we would still get some credit for being in orbit at all. Minimum success meant that you would get globally distributed science data from at least two out of three spacecraft science instruments for a minimum of 180 days from the planned science orbit or from some equal time from an unknown orbit. The rest of the success criteria were the same as for the complete success of the mission. The reason it's important to lay out the mission success so explicitly and to do it well in advance of the launch is to prevent second-guessing, post hoc analysis, and the kind of what-if exercises that are so common in our society.

By mid-March 2001 the spacecraft team at JPL and Lockheed Martin and the Mars team at HQ all felt that things were moving along reasonably well. The issues that had been identified earlier in 2000 were being addressed, and it seemed as if, after a rough beginning with so many concerns that had to be addressed, we were on a path to launch in April 2001 with no more than the usual types of issues that have to be addressed for any launch. Of course, one must never discount Murphy's law (if anything can go wrong, it will).

On March 13, I received a call from Firouz Naderi and the head of all JPL flight projects, Tom Gavin. They reported to me that an electronic part called a field-programmable gate array, an FPGA, had failed in a test for another mission called SIRTF, standing for Space Infrared Telescope Facility, a space observatory then under construction. The FPGA was a very common electronic part throughout the industry and was included in the construction of the *Odyssey* electronics box. This was the first notice I'd received of a problem and it would not be resolved until right before the launch of the mission. We were only three weeks away from the nominal first launch opportunity on April 7. Pressure on all concerned skyrocketed.

JPL leapt into action with all of the in-house experts it had at its disposal. They began an aggressive round-the-clock test program. These parts were common, not only to the SIRTF mission, but also to the Genesis mission that would launch a little bit later, as well as our *Odyssey* mission. However, given the sensitivity of our program and NASA's desire to avoid failures, it was clear that having only an in-house JPL team evaluating the situation was not sufficient. An independent assessment team for the FPGA problem was quickly assembled and began their own examination of the problem.

In the meantime, George Pace at JPL and the rest of the team were dealing with the final disposition of the other problems that had been identified. There were a sufficient number of these that a special mission-risk briefing was held on March 27, 2001. In a presentation, Pace reviewed all of the recommendations that had come out of the failure of the two previous missions and reported on what had been done. He noted that there was an external red team that had been put together in May 2000 that had followed all these problems and checked on the solutions being proposed by the project. A lot of other risk reduction and testing had been instituted as well.

The key actions accomplished to address problems that were first noted in the spring of 2000 and continued to be found through the spring of 2001 were things such as additional staff added at both JPL and Lockheed Martin to be certain there was an adequate level of both review and performance; both English and metric units labeled on all drawings, with a second level of documentation to be sure that the problem that caused the loss of the MPL didn't happen again; and independent verification of all the critical analyses and mission-critical parameters. In the press conference that occurred right before launch, George Pace noted that there were perhaps 10,000 separate parameters or functions that all had to be entered or performed correctly in order for the mission to be successful. Perhaps this is why they call it rocket science.

To address the problem that likely caused the loss of the Mars Observer mission, the project had, in the end, added the check valves to the propulsion system. The team also changed the sequence that would occur at the time of Mars insertion to allow for recovery from any type of onboard electronic upsets. They instituted a whole series of other changes to ensure the spacecraft had a robust orbit insertion. I personally required a series of power life-cycle tests to ensure that the solid-state power amplifiers would, indeed, stand up to the cycling that was being proposed. At a management level, they fixed some of the unclear or vague lines of responsibility and reporting and made sure that the development people who had designed the spacecraft would be part of the operations team, at least initially.

From this point, March 27, on, all the problems from the faster, better, cheaper era seemed to have been resolved and closed. The project noted that a total of 144 red team actions had been identified, and they had closed 142 of them. The two items that caused a disagreement between the project and the red team were taken up through the JPL Governing Program Management Council and eventually closed with the concurrence of the JPL director.

Even with all this effort, at the end of Pace's presentation on March 27 was a set of charts called residual risk items. The first one was the FPGA problem. Although the project felt that there was little residual risk because of the extensive testing that had been done, this opinion was not shared by everyone on the independent assessment team.

Just to make life interesting, while grappling with the problem of the FPGA, I also had to deal with an unexpected health problem. In early March I began having a pain in my jaw as if I were experiencing a toothache or maybe had fallen and not realized it. I went to the doctor and discovered that I was having an outbreak of shingles. If you've had chickenpox as a kid, as I did, it seems the virus lives in your nervous system forever. With enough stress on your system, the virus can make another appearance. My stress had let loose the virus on the nerve running under my jaw at the side of my face and put me down for well over a week right as we were attempting to prepare for the *Odyssey* launch. I managed to recover and jumped back into the fray, which by then was boiling over with new tests, measurements, and multiple review teams.

At this point, quality assurance, reliability, and test engineers at Lockheed Martin in Denver, at JPL, and at NASA HQ as well as at other NASA centers, particularly Goddard, were all pressing ahead with a variety of analyses, tests, and measurements to see if they could determine where the problem came from and whether or not it was a systemic problem throughout the entire batch of FPGAs.

To make sense of all of these various opinions, tests, and measurements, we decided to convene another independent assessment team, which was formed on Monday, April 2. This team included a chair and vice chair from the Aerospace Corporation, a company that is primarily concerned with helping the U.S. Air Force understand, verify, and validate their missions and launch vehicles. The company's reputation is sterling. Another member came from NASA's Goddard Space Flight Center, and the final person came from Lockheed Martin's Sandia National Laboratories. All these people were considered to be experts in the field of electronic parts and had a great deal of experience with FPGAs.

The objectives of the review and analysis by this independent team were to provide expert opinion on the validity of the analysis results, suggest possible failure mechanisms for the FPGA problems, and assess the implications of a failure on the *Odyssey* flight. The review consisted of presentations by Lockheed Martin and JPL personnel, questions by the review panel and other attendees, and a mechanical testing lab visit. With a launch date of April 7, tension was building hour by hour.

There is a month's worth of activity leading up to even the most straightforward of launches. For the *Odyssey* launch, we had to decide whether or not to press ahead with loading the liquid oxygen and liquid hydrogen fuel. We had to decide this on April 3. If we didn't go ahead, we would lose our spot in the launch lineup and would have to call off this opportunity. As of April 3, neither JPL nor Lockheed Martin had been able to determine the root cause of the failure of that FPGA. What we did know was that the SIRTF mission had had one failure and, in reviewing the records, we found that the first time the *Odyssey* electronics were turned on with the 42 FPGAs, there was a failure of one of them. The failed unit was replaced, and after 1,000 hours of testing no further failures had been observed.

The flow of these parts went like this: The manufacturer had sent a large supply of FPGAs to Lockheed Martin in Denver for use in all of their missions. Of those pieces, 218 units had been parceled out to several different programs including *Odyssey*, SIRTF, Genesis, and the Mars 2001 lander that was still in storage. SIRTF had received 38 units and had had one failure. The only other failure was the initial start-up of the Mars *Odyssey* electronics.

We decided to go ahead with the tanking, the flow of liquid hydrogen and oxygen into the rocket, in anticipation that we would be able to launch by April 7.

The independent assessment team delivered its report to me on April 5. In that report they went through their analysis of all the different failure possibilities, recommending that if it could be shown that there was a systemic problem, the parts all be replaced. They did not know if there was a systemic problem, though. In their report, they called out six key findings: Two parts out of 218 from the flight lot failed at Lockheed Martin, despite having high reliability space qualification parts screening and handling processes in place; failure analysis at Lockheed Martin and the manufacturer has not identified root causes of these failures; the failure rate calculated by the manufacturer is below the actual data; there is no evidence of a fundamental reliability problem with this particular FPGA, but the number for the failure rate for parts, 2/218, is significantly higher than expected; circuit design errors have been identified in the FPGA, but the likelihood that they are the root cause is believed to be low; and, finally, that several possible mechanisms have been suggested for each failure, but until the mechanisms for these failures are identified with high confidence, the use of these parts entails an unknown level of risk.

The team concluded that there were several potential failure mechanisms and that, of these mechanisms, a number would have benign consequences. However, other potential failure mechanisms could have much more dire consequences. The final sentence of their report was indicative of the confusion everyone felt: "As there is no simple way to determine the likely failure mechanisms with reasonable confidence before the *Odyssey* launch, it is not possible for the independent assessment team to recommend a disposition for the other parts on board in the *Odyssey* spacecraft as to sound engineering principles." So in other words, they were telling me, you're on your own, buddy.

The same day the independent assessment team delivered their report, JPL presented me with their analysis of the most likely cause of the problem and the residual risk to Mars *Odyssey*. They considered the most likely cause of the *Odyssey* FPGA failure to be some kind of electrostatic discharge or handling problem in the lab as the parts were being assembled. They went on to say that because there had been more than 1,000 test hours with no anomaly since that initial problem, the residual risk to Mars *Odyssey* was low.

To further complicate matters, one of the members of the independent assessment team chose to write a minority report. In this minority report, the expert stated that he believed there was, in fact, some type of a systemic problem in these FPGAs. He thought that there could be some type of a flaw in the basic silicon wafer from which these parts were made. He went on to say that it was conceivable that the launch vibrations would cause this flaw to expand and enlarge. It would result in the shorting out and failure of the entire FPGA device. If such a failure occurred, it would certainly compromise the mission if not cause an outright failure in the electronics of the *Odyssey* spacecraft. This would, of course, mean another failed NASA Mars mission.

After the failures of MCO and MPL, one of the findings from the Tom Young committee was that staff engineers thought they could not speak up when they saw a significant risk. Therefore, the JPL management and my team felt obligated to go the extra mile to hear any minority or alternative opinion to the prevailing assessment. I believed that we had to do this with the FPGA problem. On April 6, I asked Tom Gavin at JPL to have the project people, the reliability engineers, and the other experts at JPL review the independent assessment team findings and the minority report one more time.

April 7 dawned clear and sunny. It looked like a very good day for a launch, no mean achievement in a place where sudden storms, lightning strikes, unpredictable winds, and fluke showers routinely scrub launches. The launch was scheduled for 11:02 a.m. that day. After that launch date there were others over a span of 20 days, but 20 days was certainly not enough time to replace some 42 electronic parts buried deep inside a spacecraft that was mounted on top of a Delta II launch vehicle.

One of the unusual bits of protocol surrounding a launch is the seating assignment in the Mission Director's Center (MDC). As I came to learn, the political jockeying for the seats was almost as complex as the stories of designing the negotiating table at the end of the Korean and Vietnam wars. In the case of *Odyssey*, I was seated at a console with the designation HQ1, which meant that I was the ranking NASA Headquarters official sitting down at the station where the "go for launch" is given. Much as is shown in the movies, there really are a series of people who are asked whether or not they can give a thumbs up, or a "go for launch." This group includes the launch vehicle experts and the range safety people. It includes the tracking and communications stations and the spacecraft manager. It includes a number of other backroom engineering people. This time it included me.

By early morning of April 7 the tension surrounding the FPGA problem was continuing. Ed Weiler was over at the press site. His deputy, Earle Huckins, was in the Mission Director's Center but seated back in what they called the "Fishbowl," which is an observation area. By eight o'clock that morning most of us were already at the MDC. At 8:20 a.m. Tom Gavin handed me a document that I'd first seen two days before, the Mars *Odyssey* FPGA independent team assessment, and wrote on the lower left-hand portion of the package: "Following receipt of the Independent Assessment Team final report, there are no changes to the conclusion of the JPL position cited herein. Tom Gavin, April 7, 01 8:20 EDT."

By nine o'clock, the countdown was well along for an eleven o'clock launch, and there was one more important document to be signed. It is called the Certification of Flight Readiness (COFR) and must be signed by the launch operations people, project people, and NASA HQ; signing it basically commits the parties to go ahead with the launch unless there is some last-minute problem.

I went to Earle Huckins and asked him how he wanted to handle this entire FPGA problem. Earle said, "This is your mission, Scott. You make the decision, and you sign this COFR if you think it's the right thing to do. Whatever you decide the rest of Headquarters will stand behind you." I have rarely felt so truly alone. At the same time, I also knew that I could trust Earle to back me up regardless.

I thought about a lot of things over the next few minutes—the years of work leading up to this point, a similar situation three years earlier when the first use of the Athena II rocket was to launch Lunar Prospector, and the fate of the Mars Exploration Program riding on this next launch.

I had some experience with electronic materials, having spent six years at a national laboratory and five more years in industry working with semiconductors. Looking at the reports I had in my hands and thinking about the minority opinion, my decision was that, in this particular case, the majority was correct. After 1,000 hours of testing, no problems were found. The type of failure observed in the FPGA did not look like it was caused by a crystalline defect in the silicon. Sometimes, you just have to play to win. I signed the COFR.

If the lone dissenter was correct, of course, we might not hear from the spacecraft at all, so that first signal of health was vitally important to everyone and especially to me. The countdown continued. It was tradition for the HQ representative who was responsible for the entire mission to be one of the very last people polled. All the other stations reported "go for launch." When it got to me, I said "go for launch."

Figure 14. Hubbard and his wife, Susan, just after the successful launch of the *Odyssey* spacecraft on April 7, 2001.

The launch was smooth. We watched the rocket ascend, and we saw a stage separation. Then we started watching the telemetry, the engineering signals from the spacecraft that told us how it was managing. We knew that it would be some minutes before the fairing, the top of the rocket that is in two pieces and holds the spacecraft within its grasp, opened and the spacecraft was spun up and released. The tension was still thick. I knew that Dan Goldin, the administrator, was both watching the launch on NASA Select TV and listening in on the private line.

Finally, after what seemed an eternity, the signal came through. The spacecraft was alive and well and headed for Mars. The blood that had been thundering through my veins may finally have ebbed a bit. We were on our way back from what had been such a terrible set of disappointments in late 1999. Of course, still to come were the cruise phase and then the critical Mars orbit insertion, but at that point we had a new program supported by the administration, a new team, a new way of working together, and a new mission launched to the red planet.

I walked out of the Mission Control Center and saw my wife, who had been sitting up in the Fishbowl. We embraced each other and both began to weep. It was a release of a year's worth of tension, pressure, effort, travel, negotiations, battle, and ultimately what felt like success. A reporter standing nearby had her photographer snap a picture of Susan and me embracing after the successful launch of Mars *Odyssey*.

I felt as though I had been in an altered state for a year, working toward a successful outcome for a complex interaction of engineering, science, international diplomacy, project and program management, and politics. It is really only now, after a decade, that it's possible to see how successful this entire architecture effort has been.

CHAPTER ELEVEN

The Queue

The ticking of all the clocks had been deafening: The federal budget cycle clock, which is inexorable and unforgiving; nature's 26-month cycle for launches to Mars, even more relentless and unchangeable; and the incompressible time it takes to build special-purpose spacecraft all weighed heavy and hot on the new Mars Program team. In this supercharged environment, we had had to make a new science-driven Mars Exploration Program that was much more than a collection of missions. It was a juggling act of immense proportions with international expectations from the suspicious French, the volatile Italians, and the newly sales-minded Russians. There was a science community, at once dispirited and hoping for new leadership. The bipolar love-hate relationship between the science community and JPL was in full swing. Managers and engineers didn't know where to turn—bury themselves in process or depend on a few good people.

Firouz Naderi and I had developed a plan we called program systems engineering that was part intuition, part cost analysis, part mission design, and part technology assessment. Many, many groups had opinions, ranging from the Tom Young committee who had reviewed the twin failures to the National Academy Committee, the NASA Advisory Committee, and on and on.

My task had been to fix it and that had included getting all the various players, from NASA HQ types and JPL to government contractors and Congress, to buy in to the plans we were proposing. In a very real sense, the primary deliverable for all our hard work was the mission queue, the list of missions we proposed to fly over the next ten years to advance our understanding of Mars and prove that it was worth spending several billion dollars to do that. I confess I am quite proud of the list we developed and, having watched as things have played out in the last ten years, I think I can safely say I am justified in that pride.

Not only have the missions successfully orbited and landed, but the scientific results have been nothing short of spectacular. One early intriguing result was documentation in photos of what appear to be "recent" water runoff gullies on Mars. While this initial report of water-induced gullies has been subject to alternate explanations—which is the way of science—we now have an abundance of beautiful data from orbit and from the surface of the red planet that confirm the long presence of very salty water in large areas.

Here is the list of science missions that comprise our queue:

Mars Global Surveyor

The first mission we included in the program systems engineering for a new queue was Mars Global Surveyor (MGS). Although this mission was actually launched well before I got to Washington, the importance of MGS in science data acquisition, landing-site selection, and communications demanded that we always keep MGS in mind. To the public, orbiter missions are rarely as compelling as landed missions. They are not as readily anthropomorphized. They don't look like pets or even strange creatures crawling around alien landscapes. They are boring, boxy shapes that we never see after launch. They are literally distant, circling above their targets, seemingly monotonously. They are inherently anticlimactic. Once in a successful orbit, they tend to stay put, requiring only routine maintenance burns—small adjustments to keep them in position.

To a scientist, though, orbiter missions are the life's blood of planetary exploration. They provide enormous streams of data—images, and spectra, of a global nature that put entire planets into context. MGS was a splendid orbiting mission. From its circular polar orbit, it mapped Mars as the planet turned beneath it, creating a full pass every 12 hours or so.

MGS was equipped with a solid suite of instruments that provided a complete, high-resolution picture of Mars for the first time. As always with NASA, each instrument became known by its acronym, a habit that, while often confusing or even irritating to the public, speeds communication and in a small way helps with team building within the science community. The Mars Orbiter Camera, the MOC, took more than 240,000 images of

Mars in its nearly ten-year lifetime. MOLA, the Mars Orbiter Laser Altimeter, provided telling altitude data to help interpret those images. TES, the thermal emission spectrometer, provided fascinating insight into mineral composition. The MAG/ER, a magnetometer and electron reflectometer pair of instruments routinely used to map planetary magnetic fields, helps define a planet's core and told us Mars's magnetic field had vanished as the core cooled, but that magnetic "stripe" remnants remained. The USO/RS, ultrastable oscillator for Doppler measurements, essentially a highly refined clock, was a necessary part of accurately positioning and coordinating the other science data. Last but not least was the MR, Mars Relay, a signal receiver that was required for us to stay in touch.

MGS has also dispelled myths. Some readers of these pages may recall a silly controversy about a Viking image that looked enough like a "face" on Mars that it even prompted the plot of a not very ingenious Hollywood movie, *Mission to Mars*, that featured the location as a kind of depot of an advanced civilization, among other things. MGS imaged the area, and it was immediately revealed to be a simple mesa with a collection of large rocks and shadows.

MGS did its job, and did it well for nearly a decade, long beyond its nominal primary mission. Sadly, as yet again another testament to just how challenging Mars exploration can be, MGS was eventually lost to us, probably prematurely and probably due to human error. A data entry mistake may have caused the craft to orient itself incorrectly and overheat a motor, which resulted in a cascade of events that ultimately caused the spacecraft to run out of power.

Mars *Odyssey*

Number one completely on my watch was Mars *Odyssey*, named with great affection and respect for Arthur C. Clarke's *2001: A Space Odyssey*. *Odyssey* is an orbiter that launched on April 7, 2001, on a Delta II rocket from Cape Canaveral Air Force Station and successfully reached Mars orbit on October 24 that year. *Odyssey*'s three primary instruments are THEMIS, the thermal emission imaging system; the gamma-ray spectrometer (GRS), which includes the high-energy neutron detector (HEND), provided by Russia; and MARIE, the Mars Radiation Environment Experiment. On May 28, 2002, NASA reported that *Odyssey*'s GRS had detected large amounts of hydrogen, a sign that there must be ice lying within at least a meter of the planet's surface. We had truly begun to follow the water.

Figure 3 in chapter 3 shows the incredible discovery of the vast amounts of water ice in the first meter (3 feet) of Martian soil—greater than 60 percent by mass fraction (more than 80 percent by volume) at the poles and very large elsewhere.

Like many, indeed most, successful NASA missions, Mars *Odyssey*'s primary mission was completed by August 2004, and yet it continues serviceably to this day. JPL, given an adequate budget and reasonable, even if modest amounts of time, does a splendid job of designing and building interplanetary spacecraft. *Odyssey*, of course, besides carrying its reconnaissance instruments, had the additional task of serving as a relay station for the large data sets that we knew would be coming. *Odyssey* has performed that task extremely well and continues to do so. As Mars exploration has progressed, the great amount of data we anticipated has become vast and will do nothing but grow exponentially in the future. Without workhorse relay stations such as *Odyssey* we might lose some of these precious data.

Sir Arthur C. Clarke is no longer with us, sadly, but his legacy is alive and well, and his namesake is doing him proud around the red planet.

Mars Exploration Rovers

The Mars Exploration Rovers—known, and seemingly loved, universally as the MER twins—were launched in June and July of 2003.

Spirit, MER A, was sent to Gusev Crater because, morphologically—that is, by appearance—Gusev looked as if it might have been a crater lake and might have held water in the past. There were what looked like runoff channels capable of filling the crater if Mars had had the wet history that some from the science community believed was the case.

Opportunity, MER B, in contrast, was sent to a region where the geochemistry seemed relevant to following the water, Meridiani Planum. Large deposits of a particular form of a mineral called hematite had been identified there by MGS. Hematite on Earth is common, and the particular form of hematite in question, gray hematite, is strongly associated with large amounts of water, which facilitate its formation here on Earth. When *Opportunity* landed on Mars, the science team was jubilant. They felt as if they had literally landed a hole in one. When the MER B airbags stopped bouncing, it became clear that *Opportunity* had rolled into a small wind-scoured crater and was looking directly at Mars bedrock—the gold standard for a geologist because it had a guaranteed context. *Opportunity* quickly

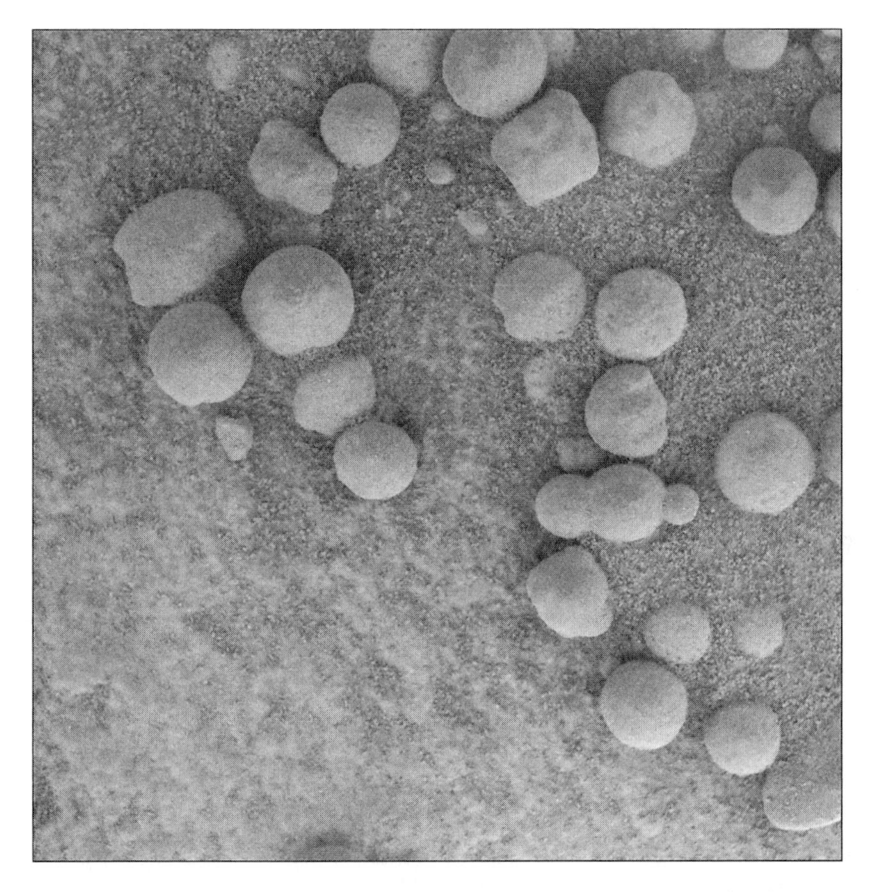

Figure 15. The *Opportunity* rover landed near a site where a water-related mineral (hematite) was detected from orbit. *Opportunity*'s microscopic camera took these close-up pictures of the "blueberry" nodules of hematite. (Courtesy NASA/JPL/Cornell University)

identified large numbers of "blueberries," small spheroids that, again, are strong indicators of water formation.

At the time of *Opportunity*'s landing, I was reunited with Gentry Lee, Firouz Naderi, and my Mars program director successor at NASA HQ, Orlando Figueroa, who took over when I left in June 2001. We were there to be part of the jubilant teams celebrating not only the hole-in-one landing but the incredible excitement of being right next to bedrock. The picture of the four of us is one of my favorites.

It's hard to be too positive about the intrepid Martian explorers, *Spirit*

Figure 16. Left to right: Orlando Figueroa, Firouz Naderi, Gentry Lee, and Hubbard celebrate on January 24, 2004, at the successful landing of the *Opportunity* rover.

and *Opportunity*. Engineering the package, though, was an incredible challenge—much akin to putting 1.01 pounds into a 1-pound sack. Figure 17 shows a somewhat humorous photo of Firouz and me next to the fully assembled *Spirit* system around Halloween.

Putting the rover with the Athena payload into the Pathfinder envelope was a triumph of packing density. To keep the cost under control, to meet the accelerated schedule, and in order not to exceed the limits of the heat shield or parachute, every cubic inch was used to the maximum extent possible. By the time the photograph was taken, it would have been difficult to slip a credit card into any gap.

By now the twin rovers have captured the hearts and minds of both the public and the science teams following them. They turned Steve Squyres into a television celebrity for a time. They showed up in McDonald's Happy Meals. They were embraced by those working with them as if they were children with fortitude, gumption, and personality. They provided bountiful "ground truth" for our orbital reconnaissance.

Each identical rover was equipped with the same set of instruments.

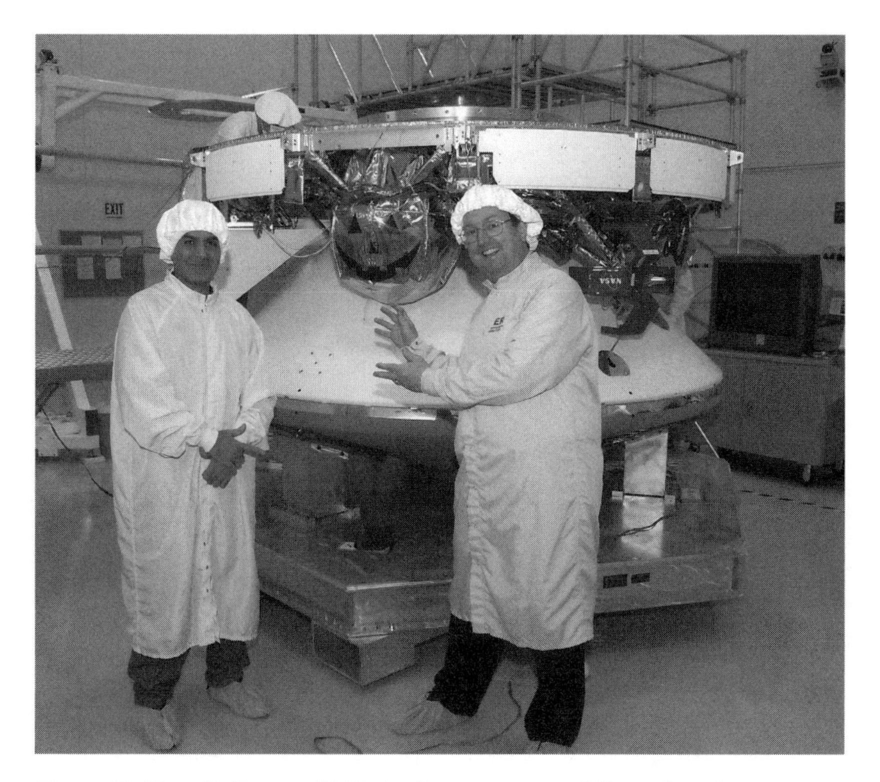

Figure 17. Near Halloween 2002, the *Spirit* rover was fully packaged in the aeroshell, heat shield, and cruise stage. Firouz Naderi and Hubbard have a bit of fun with the propulsion tank decorated for the occasion. (Courtesy NASA/JPL-Caltech)

Affixed to each rover is a Navcam, the navigation camera that provides the basic "sight" for navigation and driving. The data are monochrome with a high field of view and adequate but fairly low resolution. Then there's Pancam, a panoramic camera that examines texture, color, mineralogy, and structure of the local terrain. Mini-TES, the miniature thermal emission spectrometer, identifies promising rocks and soils for closer examination, and helps determine the processes that formed them. In addition to these instruments each rover is equipped with an arm that reaches out to get up close and personal with any interesting rock identified by the science team. The RAT, a rock abrasion tool, brushes away dust and the weathered surface of a candidate rock to expose fresh material. An alpha-particle x-ray spectrometer, APXS, does close-up analysis of the elemental abundances. A miniaturized Mössbauer spectrometer, MIMOS II, is used for close-up in-

Figure 18. Most of the Mars Program HQ team gathered for the *Spirit* landing on January 3, 2004. Left to right: Dave Lavery, Joe Parrish, Dan McCleese, Hubbard, Tim Tawney, Jim Garvin, and George Tahu.

vestigations of the mineralogy of iron-bearing rocks and soils. In addition, there are magnets for collecting magnetic dust particles, and a microscopic imager (MI) to obtain close-up, high-resolution images of rocks and soils. The cameras produce images that are 1024 x 1024 pixels, which are compressed, stored, and transmitted.

The science from the rovers has been little short of phenomenal. They were sent to Mars with the understanding that they should last roughly three months, given frigid Martian nights, totally unknown terrain, solar panels that could easily grow dim with dust, and a host of other treacherous potential obstacles both anticipated and unforeseen. In fact, *Spirit* operated for more than six years and traveled almost 8 kilometers (5 miles) before becoming stuck in the Martian soil and then losing power over the Martian winter. Remarkably, *Opportunity* is still exploring and has arrived at Endeavour Crater. When we were preparing the mission success criteria for the rovers at NASA HQ in 2001, no one, literally not one person, thought the rovers would last much longer than their design life of three months or 90 sols (Mars days). In our betting pool the latest date submitted was September 2004—about nine months after landing.

Early in their lives, operating the two rovers from Earth was a reality that in itself pushed our understanding of robotics and remote operations. Martian days are 24 hours, 39 minutes, and 35 seconds long. The early science teams mainly kept to this schedule. That is, the scientists who could, or were willing to in spite of family obligations, lived at JPL in the Mars science operations center and kept to a Martian clock. Within a week, the team members who had chosen the in-house method of following the rovers were still fresh and enthusiastic, while those who tried to juggle life on Earth with science on Mars were beginning to show signs of wear and tear.

A suite of communications tools, developed at the Ames Research Center, proved to be invaluable in maximizing collaboration among the team members and therefore maximizing the precious roving time of the twins. These tools utilize a "smart board" with a very user friendly graphical interface and continue to allow scientists, even across the country, or indeed the world, to discuss plans with colleagues while drawing in real time on a Mars landscape to suggest a route to an interesting target. They can pull up spectra and point to emission lines that show interesting elements from a lab in Boston while the same images and pointing show up at the operations center in Pasadena. They can run simulations and discuss possibilities while the solar-powered rovers are "asleep" through the cold Martian night and be ready to charge across the landscape and start sampling as soon as the sun comes up to revive the rover.

To the great delight of the teams, the very Martian windstorms and dust devils that many feared might do in the twins actually served to scrub clean the solar panels and are one of the root causes of the longevity of the rovers—that, and brilliant and careful engineering on the part of the design and fabrication teams. Another piece of the puzzle of the long and successful lives of the rovers involves the innovative thinking of the operations teams. They learned early on in the mission that by, essentially, parking a rover at the correct angle on an appropriate slope to maximize sunlight, they could coax a rover into successfully hibernating through the long Mars winter and thus survive for another year. Of course, each operations team benefited from what the other team learned at various junctions throughout their parallel missions.

The rovers seemed to live almost charmed lives. When the *Spirit* rover developed a "gimpy wheel," out of the six on each rover, by disabling an Earth-based engineering model in a similar fashion, the team learned how to command the rover to safely continue to explore while dragging the bad wheel. In fact, that dragging wheel actually dug a deeper track than it would have if not damaged and revealed new science—deposits of pure silica that could be analyzed and that fit into the emerging water picture on Mars.

That same strategy of testing a risky maneuver on Earth and then trying it up at Mars allowed the rovers to navigate slopes and explore craters that once seemed hopelessly out of reach. As Mars imagery has improved over the long life of the rovers due to the next mission in the queue, the Mars Reconnaissance Orbiter, clues that come from above can be confirmed, corroborated, and enhanced by ground testing. The interaction of the missions is working as beautifully as we hoped it would.

Mars Reconnaissance Orbiter

As important as MGS and *Odyssey* have been to our basic understanding of Mars, the Mars Reconnaissance Orbiter (MRO), launched August 10, 2005, has set a new standard with its highly sophisticated suite of advanced remote sensing science instruments. MRO sports three cameras, including the stunning High Resolution Imaging Science Experiment known as HiRISE. The secrets revealed are compelling. These high-resolution images are detailed enough to show changes such as wind erosion, avalanches, and water melt and freeze, the very processes that define the planet. The second camera assists HiRISE by establishing that all-important context and is named, fittingly, context camera (CTX). The final imager on MRO is the Mars Color Imager (MARCI), which helps characterize variations in climate. MARCI also observes dust storms and changes in the polar cap in five visible bands, plus it detects variations in ozone, dust, and carbon dioxide in the atmosphere at ultraviolet wavelengths.

Spectrometers provide emission lines to researchers, and while the public may not find them the most user friendly of science instruments, to a scientist they bring precise and critical information that cannot be gleaned in any other way. CRISM, the compact reconnaissance imaging spectrometer for Mars, has provided insights on the clays and clay-like minerals called phyllosilicates, and other hydrated silicates that form in wet environments on the surface or underground. Recently, CRISM data divulged that the northern reaches of Mars show ancient deposits of these mineralogical tracers of water, so the data are helping to refine the water picture of Martian history.

These high-resolution cameras have also served another important *program* function: landing-site selection. As we have been finding all along, one of the key elements of a true program is to have all the projects con-

Figure 19. MRO captured an avalanche of fine-grained ice and dust, possibly including large blocks of water ice, in process on Mars in February 2008. This was an amazing example of how dynamic the red planet is today. (Courtesy NASA/JPL-Caltech/University of Arizona)

tribute to the overall decades-long goals, not only the immediate mission objectives. This capability was critical in the selection of a scientifically interesting, yet safe landing site for the 2007 *Phoenix* mission. A few months before *Phoenix* landed, MRO (sometimes dubbed "Mr. O" on T-shirts) took new high-resolution pictures of the proposed site, where *Odyssey* had detected high concentrations of water ice. These images clearly showed a field strewn with large rocks—big enough to damage *Phoenix*'s landing legs. Comparisons of other sites also exhibiting large amounts of ice were quickly made, resulting in a final course maneuver that placed *Phoenix* in a safe and—as we will soon see—scientifically rich zone.

The radiometer, Mars Climate Sounder (MCS), has measured temperature, humidity, and dust content of the Martian atmosphere, expanding our understanding of the system that is Mars. Mars has clouds, weather, and climate.

The shallow radar sounder (ShaRad), a contribution of the Italian space agency, plumbs Mars's subsurface to a depth of 15 meters (almost 50 feet). Insights into layers of water ice at the north pole have been particularly significant in following the water on Mars. Between ShaRad and a deep sounder on Mars Express called MARSIS, we have compelling evidence for buried glaciers.

Phoenix: Mars Scout Lander

The Mars Scout Program was modeled after the popular and successful Discovery Program. These small, cost-effective, principal-investigator-led missions are intentionally highly focused as part of the overall cost-savings strategy. They tend to be shorter lived than most of NASA's strategic missions. They encourage innovation and offer opportunities for direct participation in space missions that might not otherwise be available to a broad spectrum of groups, particularly universities. This approach to completing the Mars Exploration Program has accomplished all this and more, in my opinion.

Phoenix launched on August 4, 2007, for a three-month mission to the frigid Martian north pole. In fact, it survived five months. Many teams were involved, headed by the University of Arizona under the direction of JPL. The science benefited from a partnership of universities in the United States, Canada, Switzerland, Denmark, Germany, and the United Kingdom. The Canadian Space Agency, the Finnish Meteorological Institute, Lockheed Martin Space Systems, MacDonald Dettwiler & Associates, and others played roles. It was the first mission to Mars led by a public university in NASA history.

For a relatively bargain-priced mission of $386 million at launch with the tightly focused goals of studying the history of water and habitability potential of the north pole, it carried an impressively diverse suite of instruments. This was possible in large part because the instruments had already been developed for other missions, an effective strategy seen with the rover twins. The instruments included:

- MARDI, the Mars Descent Imager, which, sadly, because of last-minute technical risk issues, was not used.
- A high-resolution surface stereo imager (SSI), the primary camera on the spacecraft.
- A robotic arm capable of digging down to a half meter below the surface and equipped with a color camera. This digging arm took samples of dirt and ice that were analyzed by other instruments on the lander. A rotating rasp tool, the icy soils acquisition device (ISAD), was used to cut into the permafrost. Soon after landing, the arm and camera peered under the lander and saw a white plate—ice exposed by the retro rockets. Together the arm and rasp exposed several small white "pebbles" that also proved to be ice. They sublimed—went directly from solid to gas—disappearing in a few Mars days.
- The thermal and evolved gas analyzer (TEGA), which was the darling of the *Phoenix* mission. Its task was ambitious. After the imager had identified something intriguing, the robotic arm scooped it up and essentially dropped it into one of eight small sample furnaces that would vaporize the samples so they could be analyzed with a mass spectrometer. The first sample stuck and wouldn't go through the sieve that was intended to control the sample as it entered the tiny furnace. Eventually, however, enough dirt and ice samples were tested to validate that the water ice detected by the 2001 *Odyssey* orbiter was real. And thus we had more ground truth.
- The microscopy, electrochemistry, and conductivity analyzer (MECA), which has a wet chemistry lab (WCL), optical and atomic force microscopes, and a thermal and electrical conductivity probe.

  With MECA, the robotic arm scooped up some soil and put it in one of four wet chemistry lab cells where water was added, and, while stirring, an array of electrochemical sensors measured a dozen dissolved ions such as sodium, magnesium, calcium, and sulfate. This provided information on the biological compatibility of the soil, both for possible indigenous microbes and for possible future Earth visitors. Measurements indicated that the surface layer contains water-soluble salts and has a pH close to 8—alkaline soil. Additional tests on soil composition revealed the presence of perchlorate, chloride, bicarbonate, magnesium, sodium potassium, calcium, and possibly sulfate.

The perchlorate (ClO_4) detection is considered to be particularly significant. It could be used for rocket fuel or as a source of oxygen for future colonists. While under certain conditions perchlorate can inhibit life, some microorganisms obtain energy from it. When mixed with water, it can greatly lower freezing points. Perchlorate may be allowing small amounts of liquid water to form on Mars today. Gullies, which are common in certain areas of Mars, may have formed from perchlorate melting ice and causing water to erode soil on steep slopes. Some have commented that the highly alkaline soil found by *Phoenix* would be great for growing asparagus—if it were just a bit warmer. A recent publication demonstrated that if organics (carbon compounds) are heated in the presence of perchlorates, the organics disappear and the residual chemicals resemble cleaning solvents. It is this discovery that, if confirmed, might finally explain the failure of the Viking Program to detect organics. The Viking instrument looking for organics heated the soil to 500 degrees Celsius (about 930 degrees Fahrenheit), a temperature high enough to have destroyed the organics if these perchlorate compounds were present.

- The thermal and electrical conductivity probe (TECP). The MECA contains a TECP that made the following measurements: soil temperature, relative humidity, thermal conductivity, electrical conductivity, dielectric permittivity, wind speed, and atmospheric temperature. The dielectric permittivity and electrical conductivity data were used to calculate moisture and salinity.
- The Meteorological Station (MET), which tracks daily weather with a wind indicator and pressure and temperature sensors, plus a LIDAR (light detection and ranging) device for sampling the number of dust particles in the air.

The surface of Mars has now been mapped and photographed by both the MGS and MRO missions. Very interesting features noted by both the MOC and the HiRISE camera are patterned surface areas, particularly near the Martian north pole. At times the ground has the shape of polygons. Similar features in both shape and size are found in terrestrial regions such as Antarctica. Antarctica's polygons are formed by repeated expansion

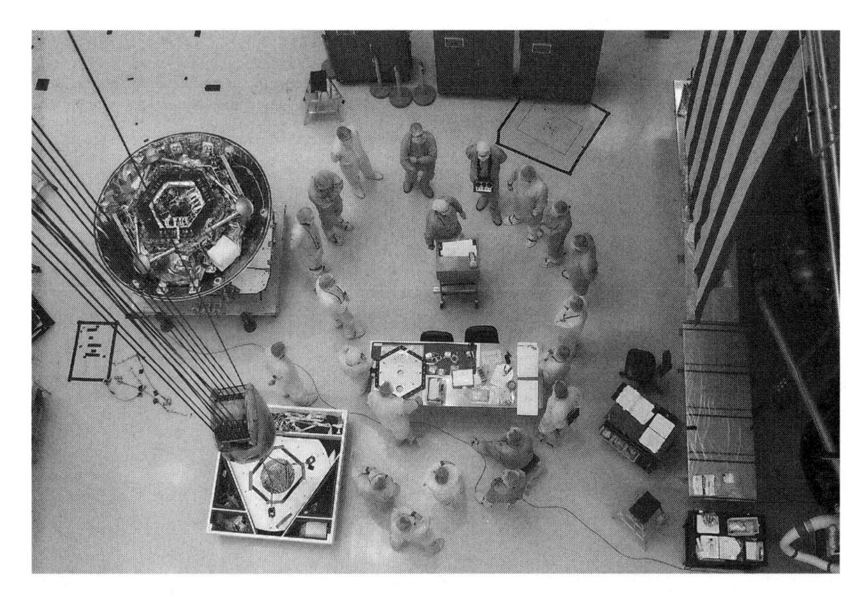

Figure 20. The *Phoenix* lander in final test in the clean room at Lockheed Martin, Denver (2006). Compare the size of this vehicle with the much larger Mars Science Laboratory rover (figure 21). (Courtesy NASA/JPL/University of Arizona/Lockheed Martin)

and contraction of the ice-and-soil mixture due to seasonal temperature changes. When dry soil falls into cracks, sand wedges are made and these wedges increase the effect. This process often results in polygonal networks of stress fractures. When the *Phoenix* lander arrived it provided close-up views of Mars's polygonal structures and was able to verify the existence of the ice-and-soil mixture. All told, *Phoenix* has significantly furthered our understanding of the ice story on Mars.

Mars 2009: Skipped; Fall 2011: Mars Science Laboratory Rover

The Mars Science Laboratory (MSL) is nothing short of a marvel. It carries the most sophisticated set of instruments ever attempted on a robotic spacecraft and it is mobile, able to carry those instruments long distances across the Martian landscape. The original concept of coupling rover capability designed to cover tens of kilometers with long-lasting radioisotope power is being realized.

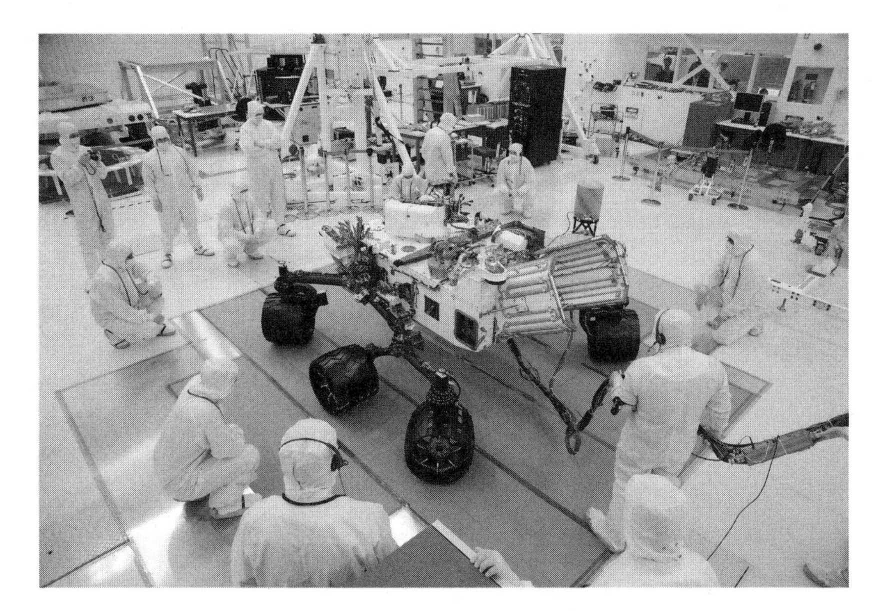

Figure 21. The Mars Science Laboratory rover in the JPL clean room during testing. At nearly 1 metric ton (2,200 pounds), this vehicle will be the most sophisticated and capable robotic lab ever placed on another world. (Courtesy NASA/JPL-Caltech)

It is not my intention here to give exhaustive descriptions of these advanced capabilities or to credit the impressive and deserving teams that have created them but merely to touch on some highlights. Such details and others are available at the NASA Mars Exploration Program website and elsewhere and may even be slightly refined as launch approaches. At present, these instruments have been selected for development or production for the mission:

- The Mars Descent Imager, MARDI, will record MSL's descent in living color at roughly 5 frames per second. This is not quite Hollywood quality, but it is surely headed in an impressive direction.
- The MSL entry, descent, and landing instrumentation (MEDLI) will measure aerothermal environments, subsurface heat shield material response, vehicle orientation, and atmospheric density for the atmospheric entry through the sensible atmosphere down to heat shield separation of the entry vehicle.
- Navigation cameras (Navcams) will tell MSL where to go and hazard-avoidance cameras (Hazcams) will help it get there safely.

- The Mars Hand Lens Imager, MAHLI, is a stand-in for one of the primary actions of a human geologist, which is to pick up a sample and examine it with a magnifier to see if it is worth pursuing further. MAHLI can take true color images at 1600 x 1200 pixels with both white and ultraviolet LED illumination for imaging in darkness or imaging fluorescence.
- For the first time, a MastCam system will give us true moving pictures, video of Mars. Two cameras will provide multiple spectra and true color imaging at 1200 x 1200 pixels and up to 10-frames-per-second, hardware-compressed, high-definition video. There are both a medium angle camera and a narrow angle camera with onboard processing capabilities.
- ChemCam, an innovative suite of remote sensing instruments, includes the first laser-induced breakdown spectroscopy (LIBS) system to be used for planetary science as well as a remote micro-imager. The LIBS instrument uses an infrared laser to target a rock or soil sample from up to 7 meters away, vaporize a small amount of it, and then collect a spectrum, giving us a precise analysis. Our French partners at CNES have played a vital role in the development of this remarkable, remotely operating chemical laboratory.
- An alpha-particle x-ray spectrometer (APXS) will irradiate samples with alpha particles and map the spectra of x-rays that are reemitted to determine the elemental composition of the samples. This instrument was developed by the Canadian Space Agency.
- CheMin stands for "chemistry and mineralogy" and is an x-ray diffraction/x-ray fluorescence instrument that will quantify minerals and the mineral structure of samples. It makes use of nano-enhanced technologies to miniaturize a system that once filled half a room but can now be flown to Mars.
- Sample Analysis at Mars (SAM) is the largest and most expensive instrument. It will analyze organics and gases from both atmospheric and solid samples. It has three subsystems involved in sample handling and preparation to collect and transport samples to three separate instruments:
    - The quadrupole mass spectrometer (QMS) detects gases sampled from the atmosphere or those released from solid samples by heating.
    - The gas chromatograph separates out individual gases from a complex mixture into its molecular components.
    - The tunable laser spectrometer (TLS) performs precision measurements of oxygen and carbon isotope ratios in carbon dioxide (CO_2)

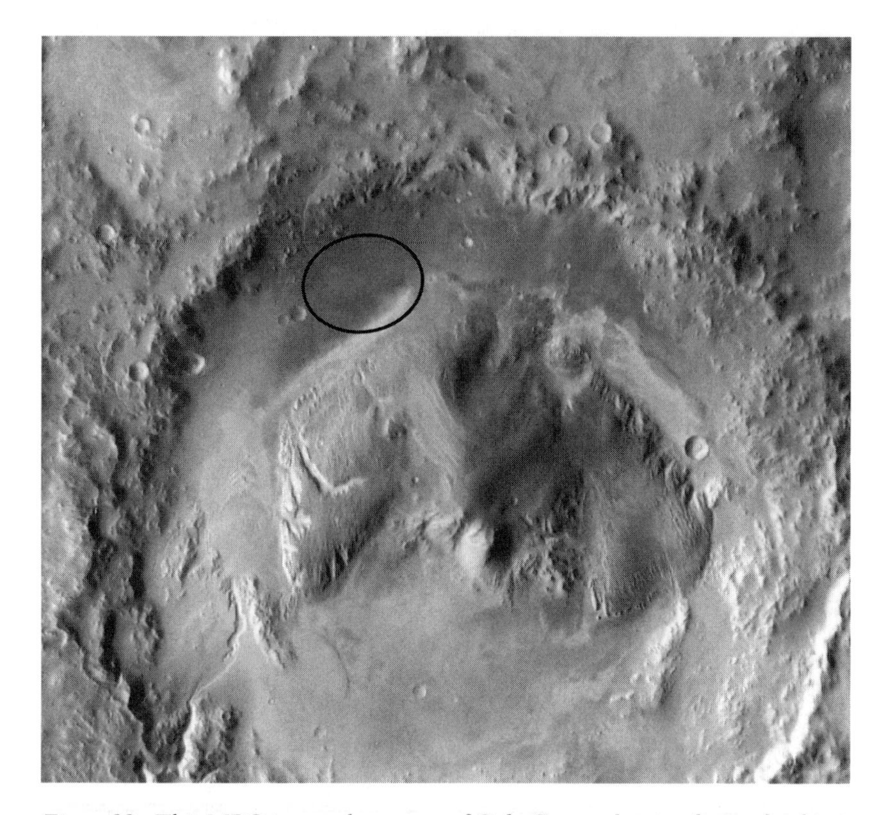

Figure 22. This MRO image shows part of Gale Crater, the site chosen for the Mars Science Laboratory rover because of the 5-kilometer (3-mile) mountain of water-related minerals in the center, right next to a safe landing area—identified by the 20-kilometer (12-mile) ellipse. (Courtesy NASA/JPL/University of Arizona)

and methane (CH_4) to distinguish between a geochemical and a biological origin.

- A radiation assessment detector (RAD) will characterize the broad spectrum of radiation at the surface to assess needs for human explorers.
- The dynamic albedo of neutrons (DAN) instrument measures hydrogen for assessing ice and water.
- The Rover Environmental Monitoring Station (REMS) is a meteorological package with an ultraviolet sensor, pressure gauges, and wind-measuring equipment to measure atmospheric pressure, humidity, wind currents and direction, air and ground temperature, and ultraviolet radiation levels.

Owing to the "ladder to Mars" philosophy of constantly building and expanding on previous missions' successes, it was indeed possible for NASA to choose a scientifically rich and geographically safe landing spot for MSL based on accumulated data from MRO. In August 2011, NASA chose Gale Crater.

We can only speculate now at how the convergence of this impressive set of instruments and mobility will advance our understanding of Mars both as a system and as a possible home for life either past, present, or future, but I assure you, I can hardly wait.

CHAPTER TWELVE

Zeroing In

As this is being written, the Mars Science Laboratory, due to launch on November 25, 2011, is the final mission of the decade for which I had responsibility. This mission was originally planned to launch one opportunity earlier, in 2009. While the change in schedule was for a host of technical reasons, the well-publicized cost increases are typical of "creeping requirements," where the science payload grew substantially from that originally envisioned ten years ago.

An element that I put in place, the Mars Scout Program, had a second project selected called MAVEN, for Mars Atmosphere and Volatile Evolution mission, set to launch in 2013. MAVEN will measure the atmosphere of Mars and try to figure out how the atmosphere disappeared and maybe a little bit about where the water went.

Figure 23 shows a filled-out ladder to Mars that has endured for ten years.

After MAVEN, the next set of missions, I believe, will shift from a theme of "follow the water" to a theme of "search for life" or, maybe more accurately, "find signs of life."

Recent data show very strong evidence of methane in the atmosphere of Mars. There are several interesting things about this discovery. One is that in the atmosphere of any planet, the lifetime of methane is short—maybe four years. This means that there must be a constant source of methane somewhere on the surface of Mars. It also mixes into the atmosphere with great rapidity, on the order of weeks, so it's very challenging to find a source, or sources. The most interesting issue is that methane can

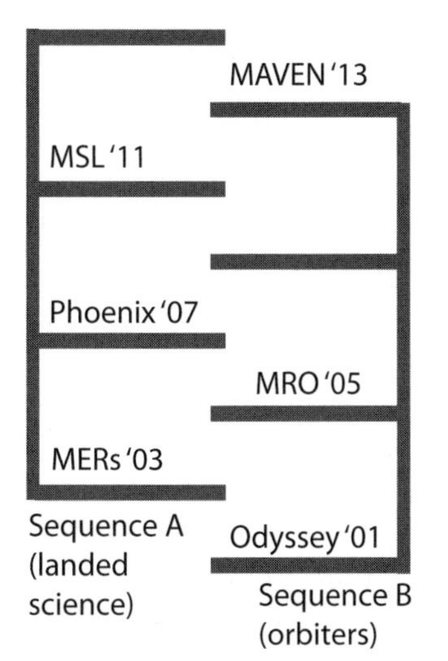

Figure 23. After ten years the success of the "ladder to Mars" as a mission queue can be seen. Early orbital results from *Odyssey* determined the landing site for *Phoenix*, just as MRO defined the target for MSL.

be produced by either of two methods. There are natural chemical and geological processes that will produce methane, but on Earth, as anyone who has ever been near a farm knows, a great deal of methane is produced by livestock from the microbes that live in their digestive systems.

So what is the source of methane on Mars? Is it possible that down below the surface there is some group of Martian microbes making a living and producing methane as a side product? Or is this simply the venting of some residual volcanic process going back to the time when the planet was much more active?

To answer these questions, a joint NASA/ESA mission is planned for the year 2016. This mission, labeled the Trace Gas Orbiter mission, will carry a whole suite of instruments that will look very carefully at the methane and try to pinpoint its source. By measuring the methane and related trace gases, we may be able to get a handle on whether the source of this gas is some natural geologic process or due to biology on the planet. As I write, this very exciting mission is still at the definition and design stage by the joint European and American teams, but it certainly will advance us another step on the "ladder to Mars" in seeking signs of life on the red planet.

What Have We Learned and What Will We Learn?

Where have all of the discoveries of the past decade led us? First, consider ancient life. As thrilling as it would be to discover extant life on Mars, even a verified find of fossilized life would change our view of ourselves forever. I believe the potential to discover ancient life has increased substantially with the new, powerful evidence in many different places of ancient liquid water, both on the surface and under the surface. We have found water ice (not to be confused with frozen CO_2) abundantly in the first meter (3 feet) of soil at the poles. We've found ice distributed in many places across the planet. We've seen runoff channels and gullies and the remnants of ancient lakes and rivers.

While these discoveries of water lead us to believe that past geological environments existed that have a reasonable potential to have preserved the evidence of life, we are also much further along in understanding what habitability means. We understand more about the limits of life and how life is so much more robust than we would have thought just fifteen years ago. Understanding habitability is proving to be a very effective search strategy to figure out where to go next on Mars. We now have the means to prioritize candidate sites to visit and explore, and we have reason to believe the evidence that we are seeking may be preserved and is within the reach of our exploration systems.

Modern life? Is it possible that there is some refuge beneath the surface, maybe miles beneath the surface, where there are living microbes today? We know that there is abundant water ice at the surface. We know from our radar measurements from orbit that there is strong evidence of glaciers buried beneath the Mars soil. We know on Earth in places like Antarctica that life exists in meltwater pockets within ice sheets. However, we have not yet identified Martian sites with a high potential for modern life, and the deep subsurface is not yet within our grasp. Still, the methane discovery may be a clue pointing toward a biosphere at the subsurface. That hypothesis is what makes the Trace Gas Orbiter for 2016 important. In any event, it is clear that Mars is much more diverse than previously thought.

With our scientific evidence of sites where there was once abundant water, where the records of Mars's past are likely preserved, what should be the next decade of missions that we launch?

Is It Time for Mars Sample Return?

It is my personal belief that the next decade—that is, a decade after the Trace Gas Orbiter, let us say 2016 to 2026—should be the decade of a Mars sample-return campaign. I have chosen the word "campaign" very carefully to mean a series of missions over a number of years that spread the cost, spread the risk, and allow time for the necessary technology development.

Why is now the time for Mars sample return? Such a mission has been the holy grail of the Mars science community for at least thirty years, if not longer. Enthusiasm bred by the success of the Viking missions prompted studies of a Mars sample-return mission, but the studies showed how difficult, not to mention expensive, such an undertaking would be. Mars sample return dominated the discussion about reshaping the Mars Exploration Program ten years ago, but I concluded we weren't ready. What has changed?

When I took over as the Mars Czar, there was a sample-return mission on the table. As discussed earlier, there were severe technological challenges to Mars sample return. However, arguably, the greatest challenge to a Mars sample return in the last thirty years has been neither the technology nor the cost, but rather the lack of scientific agreement on what samples to bring back, and from where, that would justify the cost and the risk.

The last ten years of missions and the new Mars Science Laboratory rover that is landing in 2012 have given us an enormously expanded view of Mars. I believe that at this point, there is a consensus in the scientific community, including the astrobiology community, that with the significant improvements in understanding Mars as a system, we can now take the next major step in exploring Mars. Compelling, high-priority sites for sample return have been identified. I believe the science is ready to go.

There are voices, of course, that wonder why we consider a sample return, even now. The answer is that not only have we enhanced our abilities to select promising sites, we have greatly advanced robotic science, especially mobility and autonomy. I have been asked many times by colleagues in the laboratory science community this question: If you spent the same amount of money you're spending to bring a sample back to Earth on instrumentation, couldn't you just send these advanced instruments to the surface of Mars? My answer is that if we spent lots more money, we could, in fact, get much more advanced *in situ* instrumentation for future Mars missions. There is no doubt about that at all. There are, however, three enduring and solid reasons why a Mars sample return is required. The first is that whenever you prepare a scientific payload to go to another world, you have to make many assumptions in advance about what you will find and what the right tool will be to make the necessary measurements. You won't have the chance to go to a laboratory next door, or the lab down the street, or the colleague in another city, and revisit your assumptions. The ability to follow up on measurements added to the ability to use the most advanced instrumentation available, including new advances that come along after you have launched or even returned, is a critical reason why a sample return is so much more powerful than even the best *in situ* instruments.

Second, we still don't have the ability to shrink some laboratory instruments down to the size necessary to take them to Mars or, when we do, we compromise them in various ways or are forced to make decisions based on our best guesses. Today, if you want to know the absolute age of a Mars rock, you need to use a particle accelerator to irradiate the rock and then, by measuring how the radioactivity decays, determine how old it is. No one has yet come up with a way to send a particle accelerator to the surface of Mars. Remember, getting even a single ounce to the surface of Mars requires large launch vehicles and a spacecraft with limited power and very limited volume.

There is a final reason why a sample return is so much more powerful than all the instruments you could send to the surface: With even modest samples, you can share slices among different laboratories, bringing different investigators and different approaches to measuring the rocks. The strategy of bringing back the rocks from the Moon and sharing those samples among many different investigators with instrumentation that has evolved over the years has proven to be a most powerful way to learn about the Moon. The same will be true with Mars.

While agreement in science is never absolute, the scientific community has now reached near consensus that, for Mars, the next greatest step in understanding the planet, its evolution, and the potential for either past or present life lies with getting samples back to Earth.

How do you go about doing this? What is required? We have a start already. Many elements of the decade of missions I directed were aimed at an ultimate sample return. Investments have been made in getting a relatively large mass to the surface of Mars, such as is being done with the Mars Science Laboratory. We have invested in a thorough scientific reconnaissance and surface measurement of Mars to understand where to go. We have developed entry, descent, and landing techniques that provide us with the ability not only to get to the surface, but then to travel where we need to go.

The remaining technological challenges remain formidable engineering developments. The first among these is the design and development of what is known as the Mars ascent vehicle (MAV). This is the rocket that the spacecraft carries all the way to the surface of Mars and which must endure daily temperature cycling of more than 100 degrees Celsius (212 degrees Fahrenheit). After perhaps as much as a year of waiting, this MAV must point to the right place in the sky and with high reliability blast off and then rendezvous with a waiting orbiter. Other challenges include keeping the samples clean so we are confident that they are really Martian, and so they don't accidentally kill us. This whole sequence of sample handling carries with it a whole host of extra technology challenges if drilling and coring are necessary to retrieve the samples from deep below the Martian surface.

The challenge of orbiting Mars and then carrying out an automated rendezvous with something perhaps the size of a basketball also represents a significant challenge to the robotic systems and the intelligent systems that must be designed. There have to be an Earth return vehicle and an Earth entry vehicle that will descend through Earth's atmosphere and safely bring our very expensive set of rocks to the surface with a less than 1 in 1 million chance of breaking open, and, as we have seen, you have to have a facility in which everyone is confident that you can safely handle these rocks and prove that they represent no threat to Earth's environment.

We are fortunate that not only has the Mars Program addressed many of these technological risks over the last ten years and has also collected the scientific data that we need, but other parts of NASA's program and even some Department of Defense programs have helped to reduce the risks. For example, the NASA Genesis mission, which selected samples of the solar wind and returned to Earth, and the NASA Stardust mission, which collected samples from the tail of a comet and returned them to Earth, both demonstrated the type of Earth entry vehicle we need for a Mars sample return. The Orbital Express mission funded by the Defense Advanced Research Projects Agency (DARPA) was extremely successful in showing how a space rendezvous of two different spacecraft can take place, and the National Institutes of Health and other such agencies have vast experience in handling dangerous materials and constructing the labs required.

This is not to say that coming up with a successful Mars sample-return campaign will be easy. There are many challenges to be addressed. However, I do think now is the time to embark on this adventure, and my personal approach would be to devise a campaign of three missions. The first element of a sample-return campaign would begin in the year 2018 with the launch of a rover to the surface of Mars that was capable of collecting samples and leaving them in a cache for the future missions. There is such a mission being studied even now. It goes by the name of the Mars Astrobiology Explorer-Cacher (MAX-C), but of course will probably change its name several times before launch.

With a 2018 launched mission storing perhaps as many as twenty samples of the size of a piece of chalk, the stage is set for the other two elements of a sample-return campaign. The launch might occur around 2022. I pick this year based on celestial mechanics as well as engineering and funding issues. That mission could well be the mission with an orbiter that would also contain an Earth entry vehicle. The spacecraft could wait patiently in orbit for as long as it took to get the third element of the campaign to the surface.

The final piece of the puzzle would be to send a lander with two critical elements to the surface of Mars in perhaps 2024 or 2026. One would be a Mars ascent vehicle (the MAV), and the other would be what is commonly referred to as a "fetch rover." The fetch rover would do exactly what the name suggests. It would go out perhaps some kilometers distant, gather up the sealed sample containers, travel back to the waiting MAV, and store the sample canisters in the nose cone. The MAV would then ignite, launch to perhaps 400 or 500 kilometers (300 miles or so) above the surface of Mars, and rendezvous with a waiting Earth return vehicle.

After a journey of seven to nine months, an Earth entry vehicle would separate from its parent spacecraft to speed through the atmosphere and land in the desert of Utah. The precious sample canisters would be collected and taken to a special laboratory set up to verify that the samples were harmless to Earth's environment.

Upon the arrival of those samples, I can imagine the science community embarking on an incredible journey of discovery, unraveling the mysteries of the evolution of Mars and detecting the fingerprints of life.

I have been interested in the question of life elsewhere in the universe since I was young boy of nine or ten growing up in a small Kentucky town and reading not only the science fiction of Isaac Asimov and Robert Heinlein, but also the science fact of cosmologists like Fred Hoyle. Being able to play a role in providing an answer to "Are we alone?" has been the dream of a lifetime.

Appendix

Timeline of the New Mars Program

1999

September 23	Mars Climate Orbiter disappears
December 3	Mars Polar Lander disappears

2000

January 7	Mars Program Independent Assessment Team (aka Tom Young committee) formed
February 25	First contact for me to consider Mars program director position
March 11	Meeting with Dan Goldin in Los Angeles
March 28	Tom Young committee report and announcement of my new role as program director
April 3	My official arrival at NASA HQ as program director
ca. April 6	Termination of 2001 lander
April 11	First presentation of my plans to Ed Weiler and Earle Huckins
April 24	Second presentation of replanning at NASA HQ (team formed)
May 4–5	Community meeting at JPL for 2003 concepts
May 12	First meeting with Steve Isakowitz at OMB
June 11–16	Travel to Italy and France (ASI and CNES)
July 13–14	Decision meeting at NASA HQ on 2003 mission
July 17	National Reseach Council Committee on Planetary and Lunar Exploration (COMPLEX)
August 10	Announcement of two rovers at press conference at NASA HQ
August 21–25	First synthesis meeting at JPL for 2005 and beyond
September 6–8	Second synthesis meeting in Washington for 2005 and beyond
September 14	Goldin new program briefing no. 1

September 25	Goldin new program briefing no. 2
September 29	Goldin new program briefing no. 3
October 6	OMB new program briefing
October 10	Mars Program Independent Assessment Team (Tom Young committee) review
October 26	New program rollout: press conference at NASA HQ
October 30	Solar System Exploration Subcommittee review
November 1	Space Science Advisory Committee review
November 29–December 1	Russia trip to the Babakin Institute, Lavochkin plant, and the Russian Federal Space Agency (Roskosmos)
December 4–6	*Odyssey* pre-ship review
Christmas	Passback discussions (OMB and Weiler)

2001

February 6	French space agency (CNES) discussions at NASA HQ
February 23	Solar System Exploration Subcommittee review
February 28	President's 2002 budget
April 7	*Odyssey* launch

Index

About the Author

Scott Hubbard is currently a professor in the Department of Aeronautics and Astronautics at Stanford University. He has been engaged in space-related research as well as program, project, and executive management for thirty-five years—including twenty years with NASA. From 2002 to 2006 Hubbard was the director of NASA's Ames Research Center. In 2003 he served full time as the sole NASA representative on the *Columbia* Accident Investigation Board (CAIB), where he directed impact testing that demonstrated the definitive physical cause of the loss of *Columbia*.

In 2000 Hubbard became NASA's first Mars program director (the "Mars Czar") and successfully restructured the entire Mars Exploration Program in the wake of mission failures. He is the founder of NASA's Astrobiology Institute, establishing it in 1998. He conceived the Mars Pathfinder mission with its airbag landing and was the manager for NASA's highly successful Lunar Prospector mission. Prior to joining NASA, Hubbard led a small start-up high-technology company in the San Francisco Bay Area and was a staff scientist at Lawrence Berkeley National Laboratory. Hubbard has received many honors, including NASA's highest award, the Distinguished Service Medal.

Professor Hubbard's research interests include the study of both human and robotic exploration of space with a particular focus on advanced technology and national policy. Hubbard has an ongoing engagement with robotic Mars missions, both as a member of National Academy of Science review groups and as a frequent consultant to NASA projects. Professor Hubbard is an expert on the emerging entrepreneurial space industry and serves as director of the Stanford Center of Excellence for Commercial Space Transportation.

Hubbard continues his more than forty-year interest in music by regularly playing guitar in a jazz group.